Advances in Mathematics Education

Advances in Mathematics Education is a marque monograph series in Mathematics Education. The book series is the first to synthesize important research advances in mathematics education, publishing high-quality monographs since 2009 which are now widely cited. Books in the series undergo several rounds of review to ensure rigor and scholarly integrity. It welcomes proposals from the community on topics of interest to the field. Each book contains original work complemented by commentary papers.

Researchers interested in guest editing a monograph should contact the series editors Gabriele Kaiser (gabriele.kaiser@uni-hamburg.de) and Bharath Sriraman (sriramanb@mso.umt.edu) for details on how to submit a proposal.

Ole Skovsmose

Critical Philosophy of Mathematics

Springer

Ole Skovsmose
Aalborg University
Aalborg, Denmark

Universidade Estadual Paulista
São Paulo, Brazil

ISSN 1869-4918 ISSN 1869-4926 (electronic)
Advances in Mathematics Education
ISBN 978-3-031-71374-3 ISBN 978-3-031-71375-0 (eBook)
https://doi.org/10.1007/978-3-031-71375-0

This Springer imprint is published by the registered company Springer Nature Switzerland AG
The registered company address is: Gewerbestrasse 11, 6330 Cham, Switzerland

Acknowledgements

Critical Philosophy of Mathematics was written during 2023 and 2024 in our flat in São Paulo. The flat is on the 11th floor with a magnificent view of the city. Due to my health situation, I have difficulties in leaving the flat, but in writing the book I have been around the world. I have, for instance, visited Göttingen, Vienna, Manchester, Athens, Alexandria, and Tahiti.

Many people have helped me during my travel, and I want to thank Helle Alrø, Daniela Alves, Denner Dias Barros, Denival Biotto, Manuella Carrijo, Brian Greer, Hans Hüttel, Jamaal Ince, Nadia Stoyanova Kennedy, David Kollosche, Priscila Coelho Lima, Renato Marcone, Fabio Alves Menescucci, Rafael Merola, Amanda Queiroz Moura, Luana Oliveira, Lucas Penteado, Miriam Godoy Penteado, and Marcos Vieira Teixeira for their many helpful comments and suggestions. I thank Peter Gates for his careful reading of and detailed commenting on the entire manuscript. I thank Rosalyn Sword for completing the language editing of the manuscript.

Miriam has been with me all the time with all her support and all her love. Without Miriam, this book would not have existed.

July 2024 Ole Skovsmose

Introduction

The process of writing this book has been different from those I adopted when writing my previous books. Until now I have assumed that I needed silence when writing, but this time I have been listening to music.

To help my writing the music needs to be instrumental. The volume of Billie Holiday's voice would leave me no space for putting my own words together. I enjoy the comfortable writing environment created by the guitar played by Jean Bonal, George Van Eps, Barney Kessell, Earl Klugh, Wes Montgomery, Joe Pass, Johnny Smith, and Martin Taylor. The saxophone played by Paul Desmond, Stan Getz, and Ben Webster offer my words to dance. I also like listening to Charley Parker, but sometimes his saxophone talks so fast that it disturbs my writing. My writing moves nicely forward when accompanied by the piano played by Sony Clark, Barry Harris, Junior Mance, Oscar Peterson, and Cedar Walton. I listen to Edmur Hebter, Elaine do Valle, and Toninho Pinheiro who form the group "3 Na Bossa". Gene Bertoncini's delicate interpretations of Antônio Carlos Jobim's music makes my thinking wider and lighter, but also deeper and darker. At times I listen to Ralf Zurmühle who transforms the piano into a philosopher getting lost in own thoughts. Guitar, saxophone, piano. Trumpets are too sharp, they cut my sentences into pieces.

The latest years I have become, to my own surprise, an early riser. I start my day writing and listening to music; it could be two hours before breakfast. After breakfast comes two more hours of writing. Most often I have my lunch sitting on the veranda of our flat letting my eyes touch the horizon. After lunch I need a siesta, then a bath, I answer e-mails, do some reading, and enjoy being with Miriam.

When giving lectures, I have often been asked: What philosophy of mathematics does critical mathematics adopt? In the book *Connecting Humans to Equations: A Reinterpretation of the Philosophy of Mathematics,* Ole Ravn and I present a four-dimensional philosophy of mathematics. Besides the ontological and epistemological dimensions, which concern the nature of mathematical entities and how we can come to know about them, we also include a sociological and an ethical dimension. These dimensions concern the social formation of mathematics and the mathematical formation of the social. My preliminary answer to the question: What philosophy of mathematics does a critical mathematics education adopt? has been to refer to Ole Ravn's and my book.

In this present book I fully concentrate on outlining a critical philosophy of mathematics that resonates with the concerns and hopes of critical mathematics education. I do so by exploring the following four issues:

First, I draw attention to the *social complexities* that characterise the period of Modernity, which embraces the idea that science and technology are engines of progress. I also draw attention to the fact that Modernity incorporates an extreme exploitation of manual workers and their families, brutal forms of colonisations, trading of slaves, and the formation of racist ideologies. Modernity is a period of contradictions; it comprises wonders as well as horrors. Mathematics is integral part of these contradictions; it is part of the social complexities of Modernity.

Second, during Modernity an *isolation and glorification* of mathematics dominated the philosophy of mathematics. The Modern philosophies of mathematics concentrated on intrinsic mathematical problems, which became analysed with tremendous technical sophistication. They glorified mathematics as being a unique expression of human knowledge towering isolated from mundane social, political, economic, and cultural affairs. They portrayed mathematics as minding its own business, and thereby installed mathematics in an ethical vacuum.

Third, I interpret mathematics as *indefinite constructions*. Mathematics is not reflecting any timeless realities, but makes part of complex human constructions that never reach any final format. Mathematics gets formed through ongoing processes of construction, demolishing, renovation, and reconstruction. All mathematical entities and theories could be different; mathematics could always find new and at times amazing applications; and mathematics might come to serve any kind of interests.

Fourth, I outline a performative interpretation of mathematics, seeing mathematics as intrinsically connected to actions. Mathematics-based actions might be benevolent, brutal, cynical, devastating, expensive, helpful, risky, and useful. Any kind of action, including mathematics-based actions, calls for ethical considerations, and this brings mathematics to face *ethical challenges*. Any mathematical practice, being in research, application, or education, needs to be brought out from the ethical vacuum.

If listening to music has done something to my writing, I am not able to judge myself. In fact, I never read again any of my books or articles after they have been published. I am always concentrated on my next writing tasks. The reader must judge whether this book has some characteristics that makes it different from my previous publications.

Summary of the Book

Critical Philosophy of Mathematics is composed of 16 chapters, grouped in four parts. The chapters are closely connected, but I have tried to write them in a way that they can be read separately. As a consequence, some of the chapters include minor repetitions.

Part I: Social Complexities—Modernity has been thought of as a period of time when Medieval ideas, religious dogmas, and superstitions were rolled back due to human rationality. The so-called scientific revolution replaced the geocentric worldview that had coagulated as a religious dogma, with a heliocentric worldview. The industrial revolution has been retold as a chapter of progress ensuring material welfare on a grand scale. The so-called big discoveries have been presented as initial steps towards a globalised world bringing us all together. The narratives about progress underwrites the modern self-congratulation.

Modernity, however, includes profound ambiguities. Accompanying the industrial revolution, one finds ruthless exploitation of workers and their families. The factories established horrendous working conditions, and workers had to live next to the polluting factories for the simple reason that walking was their only means of transport. Cruel colonisations took place during Modernity, as did the slave trade and the formulation of explicit racist worldviews. Genocides also occurred during Modernity. One is the North American Genocide, where the indigenous population were killed systematically. The Nazi Holocaust has been interpreted as a horrendous but isolated historical event, but the Holocaust including its "final solution" was made possible, technically speaking, only due to modern progress within science, technology, and management.

The modern narrative about progress is a Eurocentric narrative told for, and largely by, the bourgeoisie. Colonisation was a forerunner for globalisation, a notion which is decorated with nicely sounding connotations of shared concerns for everybody. However, globalisation also adds to the exploitation of large groups of people and includes cynical processes of social and economic exclusion. The globalised world is a world falling apart.

Ambiguities inscribed in Modernity are inscribed in mathematics as well. Mathematics plays an indispensable role in advancing our understanding of nature, but also of our domination of nature and of social relationships. Mathematics is an integral part of technologic developments including its wonders as well as its horrors.

For a critical philosophy of mathematics, it is important to consider the full social complexity within which mathematics is developed and brought into action. It is important to recognise how mathematics is involved, for the good and for the bad, in all spheres of life.

Part II: Isolation and Glorification—Modern philosophies of mathematics have ignored how mathematics is involved in social affairs. These absent-minded philosophies that I first of all have in mind are logicism, formalism, and intuitionism. In their glorification of mathematics these Three Magi have concentrated on epistemic issues and broadcast the dogma that mathematics is a pure discipline not serving any particular economic, political, or ideological interests.

Logicism tried to show that all mathematical concepts can be defined by logical concepts, and that all mathematical theorems can be derived from logical theorems. Consequently, logicism saw mathematical truths as logical tautologies and tried to establish a unified logic-mathematics axiomatics that would certify mathematical absolutism. The formalist programme aimed at eliminating intuitive elements from mathematical theory building. Intuition was seen as a breach through which empirical elements might flood into mathematics. For establishing mathematics in its proper format according to formalism, mathematics had to be formalised and systematically investigated. In the end formalism declared mathematics to be formal structures. According to intuitionism, mathematics cannot be identified with formal structures. Mathematics is constructed through mental and intuitive acts. This, however, does not leave space for any relativism. According to intuitionism, there is only one way in which mental constructions produce mathematics, and consequently mathematical truths turn absolute.

The philosophy of mathematical practice emerged from attempts to give the philosophy of mathematics a new beginning. It found that the philosophy of mathematics, in particular as elaborated by the Three Magi, suffered from a philosophic inbreeding and ignored what was taking place in the mathematical research practice. The philosophy of mathematical practice does not assume any absolutism, but sees mathematics as a construction which, like any form of knowledge construction, is fallible, corrigible, tentative, and evolving. Such observations led to the claim that mathematics is the mathematicians' practice.

I clearly acknowledge that logicism, formalism, intuitionism, and the philosophy of mathematical practice are deeply engaged in complex, logical, analytical, and mathematical issues. But all of them, and in particular the Three Magi, present mathematics as being detached from social affairs. By not considering what might be the social impact of bringing mathematics into action, these philosophies come to operate in an ethical vacuum. This leads to a glorification of mathematics as being a unique example of pure human knowledge.

For a critical philosophy of mathematics, it is important to recognise the position of established philosophies of mathematics, and at the same time to point out their limitations. A critical philosophy of mathematics tries to prevent from any isolation and glorification of mathematics as well as from any version of absolutism.

Part III: Indefinite Constructions—Absolutism is a dogma that makes any critique of mathematics superfluous: if things cannot be different, any critique becomes pointless. Absolutism vaccinates mathematics towards critique.

I argue that mathematics is constantly under construction and that it will never reach any definite or final form. There does not exist any Platonic reality about which mathematics might tell absolute truths. The mathematical reality is a constructed reality, comparable for instance to architectural constructions.

I find mathematics to be indefinite with respect to concepts, proofs, topics, applications, power, and culture. Mathematical notions might get new references; they might be swallowed up by more general notions; they might become out-dated and abandoned. What might be considered a proper mathematical proof changes over time and from one mathematical paradigm to another. The very idea of mathematical rigour is contested.

Mathematics is indefinite with respect to topics and applications. What for a period have been considered crucial mathematical issues might become put aside at the well-stocked museum for mathematical concepts, ideas, and techniques. New topics might enter the scene and come to play prominent roles. Mathematics always finds new and at times quite unexpected applications. What has been considered a pure, innocent, and harmless mathematical research area might become of tremendous value for technological, economic, or military purposes.

Mathematics and power are integrated phenomena. Mathematics contributes to the formation of new schemes of production and profit making; it makes new forms of social control possible; it can be used for algorithmic decision making; and it might contribute to military research. Mathematics might also establish new powerful ways of reading and writing the world, and in this way come to form part of a struggle for social and environmental justice.

Mathematics is indefinite with respect to culture; it can be encultured in various ways. The history of mathematics has, however, been recast according to a Eurocentric worldview that narrates mathematics as being a European achievement. This way mathematics has been colonised, which makes a decolonisation of mathematics an important critical task.

For a critical philosophy of mathematics, it is important to grasp mathematics as being indefinite with respect to concepts, proofs, topics, applications, power, and culture. Mathematical constructions are temporary, and can always be reworked, further elaborated, or maybe abandoned. This turns a critique of mathematics not only possible, but necessary.

The Part IV: Ethical Challenges—I advocate a performative interpretation of mathematics, seeing mathematics and action as integrated phenomena. This interpretation has found inspiration in a performative interpretation of language, in particular as expressed in discourse theory. A discourse is not a forthright tool for expressing insights and opinions; it also forms ways of seeing the world and of acting in the world. Like a discourse, mathematics forms our experiences, knowledge, assumptions, preconceptions, and ideologies. It forms categories for understanding as well as for misunderstanding. Mathematics forms our worldviews, and our life-worlds as well.

A justification of a performative interpretation of mathematics cannot be grounded in any general comparison of discourse and mathematics, and consequently I elaborate on examples of mathematical performatives. For instance, mathematics-based management fabricates the algorithmic worker as begin merely a cog to be incorporated in a clockwork-like production machinery. Mathematics contributes not only to the identification of social and environmental risks but also to the very formation of such risks. Mathematics itself tends to disguise that power gets acted out through mathematics by making conclusions, observations, and statements appear objective and neutral. By being brought into action, mathematics is shaping, fabricating, specifying, doing, forming, disguising, and articulating.

No action, nor any mathematics-based action, takes place in an ethical vacuum. The performative interpretation of mathematics brings any mathematical practice face to face with an ethical challenge. This challenge can be met in the mathematics classroom by engaging students in reflective inquiries. Such inquiries can concern the nature of mathematical ideas reveal that alternatives are possible with respect to mathematical concept- and theory-building. Reflective inquiries can concern economic inequalities and racist manifestations. They will never end in definite conclusions. They are of aporetic nature.

For a critical philosophy of mathematics, it is important to recognise mathematics as being performative. Mathematics makes part of the formation of our worldviews as well as of our life-worlds. This constitutes ethical challenges for any mathematical practice, being in research, application, or education. A critical philosophy of mathematics faces such challenges and recognises that aporetic situations make part of human life.

Contents

About the Author

Ole Skovsmose has a special interest in critical mathematics education. His research addresses landscape of investigation, dialogue, students' foreground, inclusive mathematics education, pedagogical imagination, mathematics in action, philosophy of mathematics education, and philosophy of mathematics. He is a professor at the Department of Education, Learning and Philosophy (now Department of Culture and Learning), Aalborg University, Denmark as well as a professor at Universidade Estadual Paulista (Unesp), Brazil. He has recently received the Freudenthal medal of ICMI for his innovative theoretical work.

Part I
Social Complexities

Chapter 1
Modernity and the Holocaust

Abstract The modern project embraces a belief in progress, a belief nourished by a trust in human rationality. It encompasses the conviction that after the scientific revolution, science found its proper form and has made ongoing advances ever since. The belief in progress also includes a technological optimism, seeing technology as a solid engine of progress. However, the modern project includes profound ambiguities. As part of Modernity, one sees brutal forms of colonisations, the development of slave trade, and the formulation of explicit racist worldviews. The industrial revolution established new forms of exploitation and oppression of workers and their families. Also during Modernity, several acts of genocide have taken place. The Holocaust is just one gruesome example. Mathematics is an integral part of all features of Modernity, and the ambiguities inscribed in Modernity might be inscribed in mathematics as well. A critical philosophy of mathematics must bring to the forefront the diversity of roles, including the horrific ones, that might be played by mathematics.

Keywords Modernity · Postmodernity · Modern project · Progress · Scientific revolution · Industrial revolution · Colonisation · Genocide · Auschwitz · Racism

The notion of Postmodernity seems to indicate that Modernity has come to an end, but this might be a dubious connotation. The notion of Postmodernity gained popularity after Jean-François Lyotard in 1979 published the original French version of *The Postmodern Condition: A Report on Knowledge* (Lyotard, 1984). Here he expresses a strong scepticism with respect to the possibility of formulating a unified scientific world conception. He finds it an illusion to search for "metanarratives". The only possibility is to try to create local narratives, or micronarratives. This, for Lyotard, is the Postmodern condition.

The notion of Postmodernity reaches out in many directions. In architecture it refers to a mixing of styles that have been considered incongruent. In art it refers to the mixing of, for instance, classic artistic expressions with that which might appear tasteless. Some have talked about Late-Modernity, Reflexive Modernity, or Liquid Modernity which seem to suggest that we still live in the Modern era, although we can perceive the end of it. Different as they are, these conceptions also demonstrate a

O. Skovsmose, *Critical Philosophy of Mathematics*, Advances in Mathematics Education, https://doi.org/10.1007/978-3-031-71375-0_1

broad agreement, namely that it make sense to talk about Modernity, either as something that has come to an end or as something that is fading out.

Never since Antiquity has the philosophy of mathematics reached such a degree of philosophical and technical sophistications as during Modernity. I find that the philosophies of mathematics as associated to Modernity have formed conceptions of mathematics, shaped research paradigms, and moulded approaches to mathematics education. I also find that precisely these modern philosophies of mathematics are deeply problematic. Despite their technical sophistications, they establish narrow-minded perspectives on mathematics and on mathematics education. I believe it is imperative to confront the modern philosophies of mathematics, and to open these up to other views on mathematics and on mathematics education.

1.1 Progress!

Some studies consider the belief in progress as being a singular modern phenomenon, while others have pointed out that sporadic expressions of such a belief can be found ever since Antiquity.[1] In all circumstances the idea of progress did not receive much attention during the European Middle Age. According to Christianity, life on earth was basically a forerunner of an eternal life, or maybe of an eternal punishment. In itself life on earth had no particular significance compared to that which might follow after. Life had to be dedicated to salvation.

The idea of progress changed the conception of life on earth. Life itself had a value in itself, and it could be improved! This idea contributed to a celebration of Modernity: it was a time of progress! We are dealing with an image of Modernity created by Modernity. Normally Modernity is presented as a European phenomenon. What I at times have referred to as the *modern project*, referring to an unfolding of all kind of progressive possibilities, constitutes part of European self-congratulation.[2] In the following, I will concentrate on the modern project as part of a European narrative.

The modern belief in progress was anticipated by a trust in human rationality. Many times, René Descartes has been referred to as the first modern philosopher. In *Discourse on Methods* published in 1637 in French and Latin, Descartes (1993) gave a clear presentation of the philosophic approach that he wanted to apply. Versions of scepticism have been formulated since Antiquity, and at the time of Descartes, scepticism had also found proponents, but he was deeply concerned. Descartes wanted to identify a solid foundation for knowledge securing truths that could not

[1] For a discussion of the notion for progress, see Bury (1955) and Nisbet (1980).

[2] The modern Euro-centrism has been questioned by the Indian author Ghosh (2016), who highlights that the Nava-Nyaya school of philosophy in Bengal contains "striking similarities to thought of early modern philosophers in Europe" (p. 128), and that the modern outlook was not spread around the world from Europe, but was rather a "global and conjectural phenomenon", which had many "iterations arising almost simultaneously in different parts of the world" (p. 129).

be doubted. But where are we to find such a foundation? Descartes did not want to assume any kind of dogmatism: truths should not be looked up in the Bible. According to him, a foundation for knowledge had to be identified in human thinking itself. As a first step in making this identification he applied a methodological doubt: everything that possibly could be doubted, should be doubted, and consequently not considered knowledge. Descartes applied a most radical form of scepticism and threw away our stock of assumed knowledge. Nothing survived, as apparently it was possible to doubt just about everything.

However, in trying to doubt everything, Descartes recognised one statement that he could not possibly doubt, namely that he was doubting and that consequently he must exist. He condensed this observation into the statement: *cogito ergo sum* (I think therefore I exist). The truth of this statement is not swept away by the tidal wave of scepticism. *Cogito ergo sum* is the rock-hard foundation upon which solid knowledge could be built. From this axiom knowledge can be developed through logical deduction, which has the quality of transferring truth: if the premises are true, then their logical consequences will be true as well. As Descartes considered his axiom to be incontrovertibly true, then all possible logical consequences would necessarily be true.

A trust in rationality is a trust in the human capacity to reach knowledge. No external source is needed for reaching true knowledge. It is not necessary to consult the Bible, the Church, or any other authority for obtaining knowledge. Human rationality is the supreme epistemic judge. Such a trust in rationality is a defining feature of the modern project and leads directly to a celebration of science.

The modern project is expressed in an exemplary way by the Enlightenment, guided by the idea that knowledge should be widely disseminated, and that social progress depends on the general enlightenment of people. In France, the Enlightenment project resulted in the publication of the *Encyclopédie,* which appeared in many volumes between 1751 and 1772, followed later by revisions and additional volumes. Denis Diderot was the only editor during the first years; later Jean le Rond d'Alembert became a co-editor. The principal aim of the *Encyclopédie* was to make all kinds of knowledge available to the public and in this way to enlighten the population.

In 1848, Karl Marx and Friedrich Engels published *Manifesto of the Communist Party*, and launched the socialist version of what progress could mean (Marx & Engels, 2010). The concept of progress formed an integral part of Auguste Comte's positivist philosophy. In the book *A General View of Positivism*, which original French version was published in 1848, Comte (2015) presented society as developing through different stages, the positivist stage being the ultimate one. In *On the Origin of Species*, Charles Darwin (1859) interpreted biological phenomena as developing progressively. With Darwin's inspiration, Herbert Spencer (1864) formulated the conception of Social Darwinism, seeing also social progress as a result of the survival of the fittest. In fact, it was Spencer who coined the expression of the "survival of the fittest". He found that not only biological, but also sociological, political, cultural, and ethical phenomena are subjected to the survival of the fittest.

From the beginning of the eighteenth century, we find in England explicit references made to material progress in farming, where new techniques became introduced, ensuring a more efficient output. The industrial revolution was celebrated by the modern project as being a necessary step for improving human welfare. Industrialisation made it possible to produce goods on a grand scale. The modern project was accompanied by a technological optimism: technology was seen as an important tool for humanity for dealing with an otherwise hostile nature. Through the use of technology, it became possible to dominate nature and in an efficient way to extract natural resources for serving human purposes.

During Modernity, mathematics was celebrated as an extremely successful science. Mathematics formed part of the modern project by promoting rationality, by contributing to scientific development, and by being an integral part of technological achievements. Mathematics became glorified as a sublime science.

1.2 Progress for Whom?

The Eurocentric glorification of Modernity might be based on an illusion, formed by a vastly selective presentation of what was taking place. The celebrated modern project might tell only a small part of a story, which also contains some gruesome chapters. Progress has been considered a glorifying feature of Modernity, but progress for whom?

Parallel with the scientific revolution one sees brutal forms of colonisations, the development of the slave trade, and the formulation of explicit racist worldviews. During the nineteenth century one finds several attempts to provide racism with a scientific underpinning. The intention was to provide a legitimisation of colonialism and the implementation of an imperialist world order.

I understand racism as including three claims. *First*, racism embraces the assumption that there exist different human races. Previous such differences were expressed in terms of "differences of blood". Such hidden biological differences became correlated with visible differences, for instance, differences in skin colour. Phrenology has also been applied for pointing out biological differences between the stipulated human races. The existence of different human races has been related to geographic structures. Different natural environments fostered different human races. Whatever was considered the cause of the formation of human races, a transgress from one race to another, for instance by means of education, was considered impossible.

Second, racism embraces the idea that some human races are superior to some other human races. The formulation of this claim received a strong input from the theory of evolution, which seemed to provide the discourse about the existence of superior and inferior human races with a scientific footing.

Third, racism embraces the idea that due to the stipulated superiorities, superior races have some rights that inferior races do not have: a superior race has the right to dominate an inferior race. This assumption formed an integral part of the ideological structures accompanying imperialism. For example, Great Britain had the right to colonise, to dominate, and to exploit nations populated by inferior people. The right to dominate could reach horrific proportions: the North American Genocide is one gruesome example of an attempt to eliminate a population for making space for what was coined as "progress". The Holocaust perpetrated by the Nazi in Germany is another gruesome example of an act of genocide of peoples deemed inferior as Jews and Gypsies.[3]

During the nineteenth century, phrenology turned into a carefully elaborated scientific discipline; it became an important component of physical anthropology. Through careful empirical studies proponents tried to link stipulated racial inferiority with measurable physical differences. Edinburgh became a centre for phrenological research; the Edinburgh Phrenological Society was established in 1820, and soon many such societies opened around Great Britain. By apparently being justified scientifically, racist views could be presented with authority at high tables at university colleges as well as in the English parliament. Racism turned imperialism into a noble act; as inferior races were not able to govern themselves in a proper way, they were better off by being governed by the British.

The industrial revolution has been celebrated as establishing an accelerated production of goods ensuring an improvement in human welfare and material progress. But also the industrial revolution is filled with ambiguities. The construction of more efficient and enormous production machinery resulted in making the development of factories necessary. Workers had to get to these factories, which developed into new structures of exploitation and oppression. The living conditions for workers and their families became horrendous. It is difficult to associate such oppressive features of Modernity with progression.[4]

That Modernity means progress is an element of modern self-deception, as the modern project includes ambiguous features. It is far from meaning progress for everybody, but rather only for a select minority, the bourgeoisie. It also means oppression and exploitation—even genocides.

1.3 Auschwitz

In *Modernity and the Holocaust*, Zygmunt Bauman (1989) exposes the ambiguity of Modernity in all its horrendous brutality. He starts outlining one common interpretation of the Holocaust, which describes the horror as a pathological event in human history. That the Holocaust could take place is due to the fact that Nazi-Germany fostered antisemitism to an extreme, not least driven forward by Adolf Hitler's own

---

[3] In Skovsmose (2023), I have characterised racism in term of the first two claims presented here, considering the third claim to be embedded in the second one. However, I have recognised that it might be more adequate, analytically speaking, to operate with three claims.

[4] The industrial revolution becomes addressed in Chap. 3.

fanatism. The Holocaust is described as a unique phenomenon, and not as part of some broader social trends. In this way the Holocaust gets inserted in parenthesis as being a *pathological* and isolated event, which is not relevant for grasping how the modern society functions under *normal* circumstances.

Bauman confronts this interpretation by outlining that one need not see the Holocaust as a pathological event but as a possibility crated by Modernity itself. He wants to show that the Holocaust became possible due to a combination of factors all of which formed an integral part of the modern project, and which have been celebrated as such. Only the combination of these factors was particular and unique. He finds that every "ingredient" of the Holocaust was normal in the sense of being "fully in keeping with everything we know about civilisation, its guiding spirit, its priorities, its immanent vision of the world – and of the proper ways to pursue human happiness together with a perfect society" (p. 8). This observation brings Bauman to consider the Holocaust as representing "the hidden possibilities of modern society" (p. 12).

Rather than looking into the mysteries of individual psychology, such as Hitler's conceptual universe, we need to "unravel social and political mechanisms" (p. 33) capable of manufacturing the ingredients of the Holocaust. Bauman points out that "the murder of the Jews was an exercise in the rational management of society" (p. 72). The Holocaust was a systematic attempt to apply "the stance, the philosophy and the precepts of applied science" (p. 72). When talking about the Holocaust, it is important to remember that the victims of the Holocaust were not only the Jewish population; the Gypsy population in Germany was partly annihilated too.[5]The Holocaust also exterminated groups of, among others, Russians, Polish, homosexual men, and people with mental or physical deficiencies.

According to the explanation of the Holocaust that Bauman is proposing, we need to look into some of the basic components of the modern project. We need to consider the development of rational bureaucracy, efficient management, engineering competences, and scientific insight. A blending of such "normal" factors made the Holocaust possible.

The design and construction of the gas chambers and of the ovens for burning the bodies were carefully elaborated engineering enterprises. The construction of the ovens in Auschwitz could draw on experiences of already existing constructions, but improvements were necessary. The "productivity" needed to be increased tremendously in order to meet the demands set by the "final solution". The engineering company Topf & Söhne from Erfurt became engaged in the construction of the ovens. The company had huge experience of constructing ovens for burning municipal waste. From the early 1900s, cremation became an accepted alternative to burial, and Topf & Söhne constructed crematoria for local authorities. It had become one of the world's largest companies for the construction of crematoria with exportation to more than 50 countries.

[5] See Rosenberg (2022) for an account of how he survived the horrors of Auschwitz.

The administration of the gas chambers was one of the important technical issues to be dealt with at the time. The chosen gas was Zyklon B, which was relatively cheap to produce. The Zyklon B was delivered in airtight tins containing material soaked with liquid cyanide; when the material became exposed to the air the cyanide would be spread around. In the early 1920s, the Zyklon was invented in Germany and been developed as a pest killer. It was designed to emit cyanide for a period of 24 hours. However, such a long period was not necessary for killing people; on average 15 minutes was sufficient. So simply dropping a basket with Zyklon B into the gas chamber would cause an inefficient waste of time. The tremendous "productivity" in Auschwitz was secured by the construction of columns from the floor of the gas chamber and up through the ceiling. Through these columns a basket with the cyanid-soaked material could be lowered down into the gas chamber from the ceiling of the crematory by SS soldiers wearing gasmasks. After the necessary time the basket could be lifted up again, and the dead bodies could be put into the ovens. The columns were made of solid wires, meaning that the gas could get disseminated into the room, while it was impossible for people in the gas chamber to interfere with the process. The Tesch & Stabenow company was main supplier of Zyklon B to Auschwitz.

One can wonder what kind of terminology was used when the possible functionality of a design of an oven was discussed. Maybe the very word "body" was never used; maybe the oven was meant for some "objects". The functionality of a design might well have been described using technical terminology, and not in terms of efficiency for completing mass murder. Bauman (1989) points out that a dehumanisation of the objects of the operation is taking place by talking about these objects in pure technical and apparently ethical neutral terms. Such dehumanisation might have taken place during the design process. It might also have taken place during many other processes that led to the Holocaust.

It was important that the ovens were designed in such a way that they could be managed by people without any particular technical qualification. The *Sonderkommando*, which was a squad of imprisoners from the camp, had to remove the bodies from the gas chambers and bring them to the ovens. The responsibility for keeping everything in good order in the barracks was also a task for the confined. However, it was not a Jewish person but a regular criminal with an "Aryan" appearance that was assigned this responsibility. The daily routines in the extermination camps did not demand many German human resources; much was taken care of by the condemned themselves.[6]

The locations of the extermination camps had to be decided upon. It was important that they became hidden away in less-populated areas: the smell of burning flesh would be less disturbing. At the same time, it was important that a camp like Auschwitz, crucial for the" final solution", became located in central positions on the railway network. Otherwise, the transportation of so many people to

---

[6]For a description of life in death camp, see, for instance Levi (1987). The commander of Auschwitz has presented his own "justification" of what he was doing, see Höss (1996).

the extermination camps would become too time consuming and too costly. The "supply lines" had to be kept as short as possible. The very location of Auschwitz was carefully considered.[7]

In the process that followed after the end of the Second World War, Fritz Sander, an engineer from Topf & Söhnen, stated: "I was a German engineer and a key member of the Topf works, and I saw it as my duty to apply my specialist knowledge [in the construction of the crematory] to help Germany to win the war, just as in wartime an aircraft construction engineer builds airplane, which also kill human beings." (Sanders, quoted after Pelt et al., 2019, p. 125). If one imagines that there had been lawsuit against people contributing to the construction of the atomic bombs, napalm weapons, chemical weapons, and biological weapons, one can easily imagine that they would defend themselves using a similar kind of argument as Fritz Sander.

The execution of the "final solution" was not a military task undertaken by a small band of conscripts. It was a huge task, necessitating technological competence, scientific insight, industrial productivity, elaborated management approaches, and bureaucratic efficiency. The construction of the gas chambers was a demanding engineering challenge. The efficient production and distribution of gas was an industrial achievement. The organisation of an efficient transportation of people to the extermination camps was an overwhelming organisational task, and the whole modern bureaucracy was brought into operation. The implementation of the "final solution" was made possible due to elements defining the modern project. The Holocaust cannot be explained (away) as being a unique and pathological historical event, but need to be recognised as an immanent possibly of Modernity.

Bauman concludes that the Holocaust became possible by bringing together some ordinary factors of Modernity. The modern society is *intrinsically capable* of enabling genocidal actions. This observation brings Bauman to the sinister conclusion that "cruelty is not committed by cruel individuals, but by ordinary men and women trying to acquit themselves well of their ordinary duties" (p. 153).[8]

Even drugged with only a diminutive dose of jingoism, physicists, chemists, biologists, computer experts, mathematicians, medical researchers, technicians, and managers might be turned into keen contributors to the further development of weapons of mass destruction. Bauman (1989) highlights that inhumanity is a social production, and consequently he also talks about the "social production of killers" (p.244). The modern project includes the possibility that of a "normal" person becomes reshaped as a killer.

[7] See Pelt et al. (2019).

[8] The observation is condensed in the poem "All There is to Know About Adolph Eichmann" by Leonard Cohen.

1.4 Modernity and Mathematics

Mathematics makes an integral part of all features of Modernity, and therefore all the ambiguities inscribed in Modernity might also be inscribed in mathematics. Mathematics is not automatically to be celebrated. Much philosophy of mathematics has presented mathematics as a pure science, minding its own business. Mathematics is portrayed as moving forward due to its own intrinsic logic, guided by problems that emerge within mathematics itself. I find this image of mathematics to be completely deceptive. *Mathematics is involved in all kind of social complexities.*

The existence of Auschwitz demonstrates in the most brutal way the dual role that can be played by managements, bureaucracy, technology, and science. Like these domains, so also mathematics operates in the open space between horrors and wonders. Mathematics does not represent any divine thinking any more than it represents a diabolic one. The modern project, as presented through an auto-description of Modernity, is oblivious to any such possibilities. In a similar way much philosophy of mathematic is oblivious to potential outrageous impacts of bringing mathematics into action.

References

Bauman, Z. (1989). *Modernity and the holocaust*. Polity Press and Blackwell Publishing.

Bury, J. B. (1955). *The idea of progress: An inquiry into its origin and growth*. Dover Publications.

Comte, A. (2015). *A general view of positivism*. Routledge.

Darwin, C. (1859). *On the origin of species*. John Murray.

Descartes, R. (1993). *Meditations on first philosophy in focus*. Routledge.

Ghosh, A. (2016). *The great derangements: Climate change and the unthinkable*. Penguin books.

Höss, R. (1996). *Death dealer: The memories of the SS Kommandant at Auschwitz*. Da Capo Press.

Levi, P. (1987). If this is a man and the truce. *Abacus*.

Lyotard, J.-F. (1984). *The postmodern condition: A report on knowledge*. Manchester University Press.

Marx, K., & Engels, F. (2010). *Manifesto of the Communist Party*. Marxist Internet Archive.

Nisbet, R. A. (1980). *History of the idea of progress*. Basic Books.

Pelt, R. J., Ferreiro, L., & Greenbam, M. (Eds.). (2019). *Auschwitz: Not long ago*. Abbeville Press Publishers.

Rosenberg, O. (2022). A Gypsy in Auschwitz. *Monoray*.

Skovsmose, O. (2023). *Critical mathematics education*. Springer.

Spencer, H. (1864). *The principle of biology* (Vol. 1). Williams and Norgate.

Chapter 2
Glorification of Science and Mathematics

Abstract Copernicus proposed a heliocentric cosmology which initiated the scientific revolution. Further important contributions came from Galilei, Kepler, and Descartes, before a coherent mathematics-mechanical worldview was presented by Newton. According to this view, nature is operating according to laws that can be formulated mathematically. This inspired the idea, strongly advocated by logical positivism, that mathematics is the language of science. This brought together the Modern glorification of mathematics, claiming mathematics ensures scientific unity, objectivity, and neutrality. This view coagulated as a paradigmatic dogma portraying science and mathematics as isolated phenomena not involved in any social or political complexities. This idea engenders an inadequate conception of science and mathematics.

Keywords Geocentric worldview · Heliocentric worldview · Scientific revolution · Mathematical-mechanical worldview · Glorification · Logical positivism · Objectivity · Neutrality

The so-called scientific revolution has been celebrated for making a definitive rupture with Medieval thinking impregnated with religious dogmatism and superstitions. The scientific revolution definitely brought about new ideas, but as we soon will come to see, such ideas developed alongside many other ideas, such as beliefs in astrology and alchemy for example.

The scientific revolution ascribed a particular role to mathematics in depicturing nature, indicated by Pythagorean philosophers claiming that "everything is numbers". Through this metaphysical proclamation they expressed that reality has a mathematical format by being composed of mathematical number units. Copernicus was inspired by the Pythagorean outlook when he formulated his mathematical description of a heliocentric universe. People who originally contributed to the scientific revolution—including Copernicus, Kepler, Descartes, Galilei, and Newton—were all ardent believers in God, and nature was considered as being created by God. That mathematics did ensure a profound insight into nature was due to the fact that God had created nature according to mathematical patterns.

O. Skovsmose, *Critical Philosophy of Mathematics*, Advances in Mathematics Education, https://doi.org/10.1007/978-3-031-71375-0_2

Mathematics became seen as part of God's master plan for the design of the world. Hence, there were good reasons for glorifying mathematics.

2.1 Dante's Inferno

In 1308, Dante Alighieri started writing the *Divine Comedy*. It was competed in 1321, the same year as Dante died.[1] Dante tells how he met Virgil, a Roman poet, and how they together started the journey that bought Dante and Vergil on a kind of sightseeing thrip through the inferno and the purgatory before Dante alone finally reached the paradiso. Virgil had to leave Dante before ascending to paradiso, as Vigil, being a pagan, could not enter paradise. Here Dante saw Beatrice again, his love who had died when very young.

The *Divine Comedy* was a provocative text. At that time, it was common to write in Latin considered the proper Christian language, but Dante wrote the *Divine Comedy* in Italian as spoken in the Florence region. That Dante had included Virgil in the comedy was also a provocation, as Virgil could only be a pagan as he died before the birth of Christ. When moving through the inferno Dante and Virgil saw several well-known people. They saw Ulysses, Judas, Muhammed, Alexander the Great, Bocca Degli Abbati from the nobility in Florence, Archbishop Ruggieri whose head Dante accidently came to kick, and Pope Nicholas III. They saw people being boiled with head down and feet up in buckets of water.

In 1588, Galileo Galilei was invited by the Florence Academy to give a couple of lectures. This was a prestigious academy, and the event was important to the 24-year-old Galilei: it could secure him a position as mathematician at the University of Pisa, the city where he was born. A presentation at the Academy was a good opportunity for Galilei, bubbling with self-confidence, to promote himself. What topic should he decide to lecture about?

Galilei decided to lecture about the position and the size of the inferno. Dante had indicated that inferno might have the shape of a cone, but not surprisingly, he was not able to give any specification with respect to its exact measures. Dante's inferno was a literary invention, but at the time it was often taking as being a reliable description. Galilei could tell that the inferno-cone had its apex placed in the centre of the earth, and that its base was a circle on the surface of the earth with Jerusalem as its centre. The Florence academy was impressed, and the year after in 1589, Galilei was appointed professor in mathematics at the University of Pisa.

[1]Only in 1455, Guttenberg made the first printed version of the Bible. The first copies of the *Divine Comedy* were handmade writings. See Dante (2003).

2.2 A Scientific Revolution

A Pythagorean with the name Philolaus (around 470 BC—around 385 BC) described a cosmological system where a "central fire" was allocated to the centre of the universe with the earth, the moon, and the sun circulating around. Aristarchus of Samos (around 310 BC—around 230 BC) was the first to suggest that the sun was central with the earth circulating around.

Such ideas were easily crushed by the massive phalanx of commonsense: just look how the sun is raising in the morning, walking across the sky during day, and setting in the evening. The geocentric worldview is inscribed in our language. We talk about a sunrise, the sun standing high on the sky, a beautiful sunset, and many people like to enjoy a sundowner.

In Antiquity the geocentric worldview got a powerful philosophic underpinning through the work of Aristotle (384 BC—322 BC). In accordance with Aristotle's metaphysics, Ptolemy (367 BC—282 BC) described the planets and the sun as moving in circles around the earth. That the movements took place in circles did not need any further explanation at the time, as it was taken as given that the natural movements of heavenly bodies were circles. In line with Pythagorean mysticism, behind the multitude of phenomena there existed a reality operating according to simple mathematical principles. If such principles could be exposed, one had depicted this reality, and this was what Ptolemy did in terms of circles. However, already at the time of Ptolemy, the number of astronomic observations was enormous and so precise that it became obvious that they did not fit completely with any simple circular movements of the planets. Much more sophisticated devices had to be introduced, such as the epicycles, which are circular movements, where the centres are also moving in circles.[2]

A strong ideological support for the geocentric worldview was found in the Bible, which states that God created the universe with the earth in its centre. This became a major part of Catholic dogmatism which could not be questioned without severe punishment. The Inquisition combated any such kind of heresy. The Inquisition had been established at the beginning of the twelfth century, and during the fifteenth and sixteenth centuries it expanded in power and brutality. Giordano Bruno (1548–1600) had stated that the universe was infinite.[3] One could think of this as being a metaphysical or philosophical speculation, but it could also be seen, as the Inquisition saw it, as being an audacious heretic idea that needed to be eliminated straight

---

[2] Ptolemy (2014) stated that the "earth does not move at all" (p. 32). He assumed that heavenly movements were circular, but was concerned about the observed anomalies. He stated: "The cause of their anomalistic appearance may raise on roughly two primary, simple hypotheses. For the motion is observed relative to the circle that is conceived of as being concentric with the universe [. . .] such that our sight does not appreciable differ from its center. Hence one must suppose that they make motions that are uniform either on circles non-concentric with the universe, or else on circles that are concentric with it but not upon them simply but, rather, upon other cycles carried by them and called 'epicycles'." (p. 95)

[3] See Bruno (2014).

away. If the universe were to be infinite, it could not have any well-defined centre, and consequently the earth could not be claimed to occupy its centre. In 1600 Bruno was burned alive on Campo de' Fiori in Rome.

By stating that everything is numbers, the Pythagoreans introduced an idea that came to play an important role in the scientific revolution. Nicolaus Copernicus (1473–1543) was deeply inspired by the idea that by means of mathematics one could come to grasp how the universe was organised. With unique mathematical skills, he investigated the ever-increasing number of astronomic observations which had made it necessary to complicate further the device of epicycles and other mathematical "tricks" for ensuring a fit between the mathematical model of the universe and the empirical observations. At that time the description of the movements of the planets had turned into a monster of geometric complexities. This monster was a direct negation of the Pythagorean idea that reality operated according to a few simple mathematical principles. Copernicus, however, recognised that a tremendous simplification could be achieved if the sun was assumed to be located at the centre of the universe.

Copernicus took as a given that the movements of heavenly bodies must be circles, which together with straight lines were considered archetypical simple mathematical entities, and consequently it was necessary for him to continue operating with epicycles, but their number was considerably reduced in comparison with the geocentric model. Copernicus' heliocentric model could be considered a model of how the universe was in fact operating, but it could also be considered as being merely a device for dealing with calendarial issues. It was a high priority for the Catholic Church to determine the holy holidays as precisely as possible. This was not a question of decision making, but had to be based on careful and correct calculations. The correct determination was a research issue of the uppermost importance, as it might ensure the eternal life in paradiso.

It seems obvious that Copernicus himself saw his heliocentric model as showing how the universe in fact was operating. Copernicus' main work *On the Revolutions of the Celestial Spheres* was originally published in Latin in 1543, the same year as Copernicus died (see Copernicus, 1995). The legend tells that he only came to hold in his hands a copy of the book on his death bed. Andreas Osiander, who was taking care of the printing, had added a preface, where he emphasised that the heliocentric model should be considered first of all as being a device for calendarial calculations. In this way he tried to prevent retaliations from the Church. However, in 1616 *On the Revolutions of the Celestial Spheres* was placed in the Index of Forbidden Books, where it remained until 1758.

2.3 Appearance and Reality

In 1623 Galileo Galilei (1564–1642) published the book *The Assayer*, which original Italian version has the title *Il Saggiatore*. In *The Assayer* Galilei (1957) pointed out that mathematics is necessary for understanding the grand book of nature. In order to read this book, one needed to "comprehend the language and read the letters in which it is composed" (p. 237). Nature is "written in the language of mathematics, and its characters are triangles, circles, and other geometrical figures, without which it is humanly impossible to understand a single word of it" (pp. 237–238).

Galilei made a distinction between appearance and reality. What we experience as natural phenomena might be very different from what actually is taking place in nature. The distinction between appearance and reality has a direct relevance for the conception of the universe. One experiences that the sun rises in the morning and that it goes down in the evening. This is how things appear to us, but the reality could be quite different. In *The Assayer*, Galilei did not engage in the conflict between the geocentric and the heliocentric world view, but the distinction between appearance and reality was an essential preparation for assuming a heliocentric worldview.

Galilei pointed out that tastes, odours, and sounds are just names without references to any real physical objects. They only refer to phenomena as they appear to us. Galilei added that if living creatures were removed from the earth there will be no tastes, odours, or sounds. But what then would be left? It was Galilei's conviction that for bringing about tastes, odours, and sounds "nothing is required in external bodies except shapes, numbers, and slow or rapid moments" (p. 276). He highlighted this point by stating that if sense organs were removed, there will be no tastes, odours, or sounds, but there will be shapes, numbers, and motions. Such entities make up the underlying reality, which stimulates our sense organs. Galilei referred to some miniscule particles that "may enter by our nostrils and strike upon some small protuberances which are the instrument of smelling" (p. 276). The sensation of a smell depends on the physical properties of the miniscule particles which might "have this or that shape", be "fast or slow", and be "numerous or few" (p. 276).

After this clarification Galilei turned to the experiences of heat. A question often raised is: What is fire? Could it be a kind of devouring substance? When one lights some dry wood, what does one then add to the wood? When one burns one's fingers, what is it precisely that causes the pain? Compared to what was common at the time, Galilei gave a completely new interpretation of the phenomenon of heat. According to Galilei:

> Those materials which produces heat in us and make us feel warmth, which are known by the general name of "fire" would [. . .] be a multitude of minute particles having certain shapes and moving with certain velocities (p. 277).

Whether we experience the heat to be pleasant or unpleasant depends on "greater or smaller speed of these particles" (p. 277). Besides such physical properties there is not "any other quality in fire corresponding to 'heat'" (p. 277)

Through his discussions of heat, Galilei prepared for claiming that a general distinction between appearance and reality is important for understanding any natural phenomenon.[4] For coming to understand nature one cannot rely on what

[4]The distinction between appearance and reality became fundamental to the empiricist outlook, and in *An Essay Concerning Human Understanding*, first published in 1689, Locke (1997) made a distinction between primary and secondary sense qualities. The primary sense qualities concerns

one immediately sees and experiences. One has to look behind the captivating scene of appearance. For grasping the basic properties of the underlying reality, mathematics is essential. This is the language for talking about shapes, numbers, and movements; this is the language in which the big book of nature is written.

2.4 A Mathematics-Mechanical Worldview

By carefully analysing the available observations of the movements of the planet Mercury, Johannes Kepler (1571–1630) realised that it gave a much better fit between observations and geometric interpretation if one considered the planetary movements to be, not circles, but ellipses. By launching this possibility, Kepler broke with a metaphysical assumption that had existed since Antiquity. We are dealing with a revolutionary step in the conception of heavenly movements.[5] Kepler framed the idea that nature operates according to some laws, and the very concept of natural law came to play a defining role in the scientific revolution.

In 1633 René Descartes (1596–1650) concluded the book *The World*, where he presented his general view on nature, as well as on more specific topics as the functioning of human eye, the nature of light, and meteors. However, Descartes did not publish the book. In 1632, Galilei had published the original Italian version of A *Dialogue Concerning the Two Chief World Systems: Ptolemaic and Copernican*, where he described a dialogue between two imagined figures Salviatti and Simplico. Salviatti argued in favour of a heliocentric worldview, while Simplico maintained a geocentric view (see Galilei, 2001). It appeared that Simplico became the winner of the discussion, but if we take a closer look at the very content of the argumentations, it appears that Salviatti presents a much more coherent argumentation than Simplico. There seems no doubt that Galilei himself sided with Salviatti. In 1633 Galilei had to face the Inquisition. He was threatened with torture, but got a milder punishment namely to stay in house arrest for the rest of his life. Descartes came to know about the process against Galilei. Descartes also sided with Salviatti, so he decided to let *The World* remain unpublished. The book was only published posthumously in 1664.

In *The World*, Descartes (1998) presented three laws of nature, according to which he claimed nature to be operating. As Descartes assumed that God is the creator of the universe, he found such laws to be imposed on nature. They were the

---

experiences of shapes, positions, and motions, while the secondary sense qualities refer to experiences of, for instance, sounds, tastes, and odours.

[5] Kepler condensed his observations in three laws according to which he found the planets to be moving. The first law states that the orbit of a planet is an ellipse with the sun located at one of its foci. The second law states that a line connecting a planet and the sun will cover equal areas during equal intervals of time. The third law states the square of a planet's orbit around the sun is proportional to the cube of the length of the longest diameter of the ellipse.

"traffic laws", institutionalised by God for regulating the movements of physical bodies:

> *Law 1*: Each particular part of matter always continues in the same state unless collision with others forces it to change its state. (See p. 25.)
>
> *Law 2*: When one of these bodies pushes another it cannot give the other any motion except by losing as much of its own motion at the same time; nor can it take away any of the other's motion unless its own is increased by the same amount. (See p. 27.)
>
> *Law 3*: When a body is moving, even if its motion [. . .] takes place along a curved line, [. . .] each of its parts individually tends always to continue moving along a straight line. (See p. 29.)

After presenting these three laws, Descartes stated, no doubt with proudness in his voice, that he tended to believe that God had not imposed any other laws on nature. Descartes found that he had been able to enumerate precisely the basis laws of nature, and that other natural laws could be derived from these three basic ones.

At the time of Descartes the mechanical clock was well known, and it appeared to be the most sublime technical construction at the time.[6] It was high-tech, and it got a metaphoric power. It demonstrated how a mechanical instrument does operate, and it turned into an exemplary illustration of how the universe was supposed to operate. It is a machine that runs like clockwork, and God is the ingenious watchmaker. He has designed the perfect clockwork, and when constructed it would move on forever on its own accord.

This conception of nature led to a strong form of determinism. By being omniscient, God would know everything that was possible to know, as he knows the laws according to which nature is operating. By simply calculating forwards or backwards, God would know the future as well as the past. For God to look at the history of the world, was like for us to watch a well-known video. We can step forwards and backwards, but always the same events will take place. There are no surprises for us. In the same way, there are no surprises for God. The very notion of being surprised does not apply to a Divine perspective.[7]

In 1687, Isaac Newton published *Philosophiæ Naturalis Principia Mathematica*. It was written in Latin, and translated into English as *The Mathematical Principles of Natural Philosophy*. Like Descartes, Newton also presented three laws of nature being of strict mechanical nature. Newton's (1999) laws are the following:

---

[6] A mechanical clock had previously been constructed in China, but during the thirteenth century it was reinvented in Europe. In 1292, Canterbury Cathedral got a clock installed. In 1430 a spring clock was constructed, and later the pendulum clock was invented. Galilei had mentioned the possibility of making such a clock, but it was Huygens who in 1657 made the proper design of the pendulum clock, as he identified the mathematical formula that relates the length of pendulum with the time of an amplitude.

[7] This determinism got a tremendous impact. Luther refused the possibility of human beings having an influence on God's decisions. In particular, it did not make sense to pay indulgence to the Catholic Church to ensure a place in paradise. The idea of predestination was further highlighted by Calvin. The future was fully determined, also who would come to enter paradise and who would go to hell.

> *Law 1*: Every body perseveres in its state of being at rest or of moving "uniformly straight forward," except insofar it is compelled to change its state by forces impressed. (p. 416)
>
> *Law 2*: A change in motion is proportional to the motive force impressed and takes place along the straight line in which that force is impressed. (p. 416)
>
> *Law 3*. To any action there is always an opposite and equal reaction; in other words, the actions of two bodies upon each other are always equal and always opposite in direction. (p. 417)

Like Descartes, Newton found the laws of nature to be imposed on nature by God. Like Descartes' laws, Newton's laws were mechanical, meaning that a consequent determinism also made part of Newton's conception of nature. However, there was a principal difference between Newton's worldview and that of Descartes. While the clockwork metaphor applies to Descartes' image of nature, it does not apply to Newton's. For a clockwork mechanism to function, the gears need to be interacting directly. In clockwork there is no interaction across distances. An isolated gear will not interact with anything. Contrary to this idea, Newton's introduced a force that operates across distances, namely *gravity*. Newton stated that any two physical bodies attract each other with a force that is proportional to the size of the two bodies, and inverse to the square of the distance between the two bodies (see Newton, 1999, p. 564). The very notion of gravity took the concept of nature beyond the clockwork metaphor. Newton's conception of nature was not material-mechanical; it is what I refer to a *mathematics-mechanical worldview*. The mechanism of nature, including the interaction across distances, can be described in mathematical terms, although not in a material-mechanical way.[8]

How did Newton get to the idea that there exists a physical attraction between physical bodies across distances? From where did he capture the notion of gravity? Newton was influenced by, for instance, Hermeticism which has its roots in Greek and Egypt mysticism. Hermeticism assumes the existence of an attraction across distance. Such ideas might have inspired Newton to acknowledge the possibility of an attraction between physical bodies even at a huge distance. However, although the very concept of gravity might include a good portion of mysticism, the notion operates extremely efficiently in bringing a mathematics-mechanical worldview together.

This worldview is a culmination of the Pythagorean conviction that behind the multitude of apparent phenomena exists a reality that operates according to simple mathematical principles. This conviction has become defining for the scientific revolution and for the modern worldview.

Like most others contributing to the scientific revolution, Galilei, Kepler, and Newton were strong believers in astrology. Due to the too common backwards reading of history, this belief has been seen as a regrettable residue of pre-scientific thinking. However, to Galilei, Kepler, and Newton astrology made integral part of the scientific worldview of the time. One can even see Newton's conception of gravity as supporting astrological ideas. Gravity was a force also operating across huge

[8]The notion of mathematics-mechanical worldview was introduced in Ravn and Skovsmose (2019).

distances, meaning that it was in accordance with the newly established scientific worldview to assume that heavenly constellations had an impact on what was taking place on earth. Newton firmly believed that God had created the universe, and this brought him to study the Bible carefully, and in particular he wanted to identify the proportions of the original Jewish temple. This temple could be seen as a symbol of the world, and Newton's hope was to show that some numerical proportions from the temple could be found in measurable proportions of the universe. This would testify God as the creator of the universe. Newton, however, did not manage to identify any such similarities.[9]

2.5 Mathematics: The Language of Science

In 1666, Newton drafted a manuscript where he developed the principal ideas of infinitesimal calculus. This was well before he in 1967 published *The Mathematical Principles of Natural Philosophy*. However, in the *Principia* he did not make use of infinitesimal calculous for formulating the laws of nature or for demonstrating some of their consequences. Instead Newton presented his technical argumentation in the form of geometric reasoning following a Euclidean pattern.

Mathematics and natural science grew together in one integrated network of concepts and theories, and calculus became indispensable for this networking. This brought mathematics and natural science together in the modern conception of science.

A most elaborate expression of this conception is found in logical positivism, which developed from the work of the Vienna Circle, formed in the beginning of the 1920s by philosophers, physicists, and mathematicians associated to the University of Vienna. They had the ambition of directing philosophic analysis into a new direction. The basic conviction of the circle was that traditional philosophic theories did not provide any profound insight, but rather were expressions of conceptual misunderstandings. Such theories would be better ignored. Philosophy had to be developed in a new format, and logical analysis should become its principal methodological approach. Inspiration had to be found in recent developments in natural science, mathematics, and logic.

The most important members of the Vienna Circle were Moritz Schlick (1886–1936), Rudolf Carnap (1891–1970), Herbert Feigl (1902–1988), Kurt Gödel (1906–1978), Hans Hahn (1879–1934), Otto Neurath (1882–1945), and Friedrich Waismann (1896–1959). The meetings in the circle were chaired by Moritz Schlick who in 1922 had been nominated as professor in *Naturphilosophie* (natural philosophy) at the University of Vienna.[10]

[9]For a discussion of Newton's religious views, see Iliffle (2017).

[10]For presentations of the Vienna Circle, see Ayer (1959), Edmonds (2020), Sigmund (2017), and Stadler (2015).

In 1929 the Circle published a manifesto with the title *Wissenschaftliche Weltauffassung: Der Wiener Kreis* (*The Scientific World View: The Vienna Circle*). The manifesto was written by Hahn et al. (n.d., internet), but most likely many more members of the Circle had contributed with comments and suggestions. However, not Schlick, who received the manifesto as a surprise gift when he, after having spent a semester in the USA, returned to Vienna. In a condensed way *The Scientific World View: The Vienna Circle* presents the outlook of the Vienna Circle, and with this manifesto the position of the Circle became broadly known, and it became further elaborated as logical positivism.

Logical positivism assigned a particular role to mathematics as being the language of science. Mathematics would help to ensure the *unity of science*. The basic idea was that if one provides a more systematic logical analysis of scientific theories, one will discover that behind apparent differences there exists a unifying structure. There does not exist any principal difference between sociology, psychology, biology, chemistry, and physics. Sociological notions can be reduced to psychological notions; psychological notions to biological notions; biological notions to chemical notions; and chemical notions to physical notions. They all belong to the grand family of science.

According to logical positivism, mathematics will ensure that *science becomes objective*. By this expression I refer to the claim that by means of mathematics one is able to captures features of reality that otherwise might be disguised. Galilei made a distinction between appearance and realty, and claimed that nature in reality is machine-like, and that this machinery can be expressed in terms of shapes, numbers, and motions. Mathematics is the language that can capture this mechanical reality. In this sense mathematics ensures scientific objectivity.

By functioning as the language of science, mathematics will, according to logical positivism, ensure that *science becomes neutral*. Mathematics does not leave any space for value-judgements; it is a language devoid of ethical components. To logical positivism it is important to ensure the neutrality of science—science could not include any political or ethical preferences. Logical positivism was formulated during a period when Nazi ideologies were becoming dominant in Germany and Austria, which also had an impact on what was considered proper science; some work became relegated as "Jewish science". Logical positivism confronted any such politisation of science.[11]

By ensuring the unity, objectivity, and neutrality of science, mathematics was assigned by logical positivism a principal role in scientific theory building. This is a most elaborated version of the modern glorification of mathematics. This glorification became part of a broadly assumed working philosophy of science, which proliferated as many members of the Vienna Cycle had to flee from Nazi-controlled Austria. Subsequently, many settled down in the USA. In 1931, Herbert Feigl had

[11] See Chap. 15, for a discussion of the principle of verification that was used by logical positivism for eliminating metaphysical elements from scientific theory building, include any kind of value judgements.

moved to the USA, where he together with Albert Blumberg published the paper "Logical Positivism: A New European Movement", which gave a short presentation of the outlook of the Circle (Blumberg & Feigl, 1931). The label "logical positivism" was introduced, and this label soon become the common name for visions on science and philosophy inspired by the Vienna Circle.

In 1968 I started studying at Copenhagen University. At that time one introductory discipline was obligatory for all students, namely *filosofikum* (*Examen philosophicum*). It dealt with some philosophical topics that were considered relevant for anybody beginning university study.[12] The more specific content of the *filosofikum* was decided upon by the leading professors in philosophy. In 1926 Jørgen Jørgensen (1894–1969) became nominated professor in philosophy at Copenhagen University. He was closely associated with the Vienna Circle, and he advocated strongly the outlook of logical positivism. This had a direct impact on how he specified the curriculum for the *filosofikum*. One of the basic books was *Indledning til Logikken og Metodelæren* (*Introduction to Logic and Methodology*). Here Jørgensen (1942) presents the basic ideas of formal logic and set theory. For me this came as a nice surprise. I had never imagined that such specific mathematical topics could be considered obligatory for everybody starting to study at a university. Only later I came to understand the reasons for these topics being included in the *filosofikum*. According to Jørgensen it was important for all students to master formal methods, being the grammar of the language of science.

I find the idea that science is unified, objective, and neutral to be a dogma that has been formed as part of the modern glorification of science, and not as an expression of an insight concerning the actual functioning of science. Throughout this book I am going to confront this dogma.

References

Ayer, A. J. (Ed.). (1959). *Logical positivism*. The Free Press.

Blumberg, A., & Feigl, H. (1931). Logical positivism. *The Journal of Philosophy, 18*(11), 281–296.

Bruno, G. (2014). *On the infinite, the universe and the worlds: Five cosmological dialogues* (Translated and introduced by Scott Gosnell). Huginn, Munnin and Co.

Copernicus, N. (1995). *One the revolutions of heavenly spheres* (C. G. Wallis, Trans.). Prometheus Books.

Dante, A. (2003). *The divine comedy: The inferno, the purgatorio, the paradiso* (J. Ciardi, Trans.). New American Library.

Descartes, R. (1998). *The world and other writings*. Cambridge University Press.

Edmonds, D. (2020). *The murder of professor Schlick: The rise and fall of the Vienna circle*. Princeton University Press.

[12]The *filosofikum* was initiated in 1675 and abolished in 1971. Originally the *filosofikum* was composed of grammar, logic, physics, ethics, arithmetic, geometry, and astronomy. In 1847 the content became redefined to deal with logic, psychology, ethics, and the history of philosophy.

Galilei, G. (1957). *Discoveries and opinions of Galileo* (Translated with an Introduction and Notes by Stillman Drake). Anchor Books.
Galilei, G. (2001). *Dialogue concerning the two chief world systems*. The Modern Library.
Hahn, H., Neurath, O., & Carnap, R. (n.d., internet). *The scientific conception of the world: The Vienna Circle*. https:the-scientific-conception-of-the-world-the-vienna-circle(rreece.github.io)
Iliffe, B. (2017). *Priest of nature: The religious world of Isaac Newton*. Oxford University Press.
Jørgensen, J. (1942). *Indledning til logikken og metodelæren*. Munksgaard.
Locke, J. (1997). *An essay concerning human understanding*. Penguin Books.
Newton, I. (1999). *The principia: Mathematical principles of natural philosophy* (B. Cohen & A. Whitman, Trans.). University of California Press.
Ptolemy, C. (2014). *The almagest: Instruction to the mathematics of the heavens* (Selections translated by Bruce M. Perry. Edited with notes by William H. Donahue). Green Lion Press.
Ravn, O., & Skovsmose, O. (2019). *Connecting humans to equations: A reinterpretation of the philosophy of mathematics*. Springer.
Sigmund, K. (2017). *Exact thinking in demented times: The Vienna Circle and the epic quest for the foundations of science*. Basic Books.
Stadler, F. (2015). *The Vienna Circle: Studies in the origins, development, and influence of logical empiricism*. Springer.

Chapter 3
Domination by Technology

Abstract According to modern self-understanding, the industrial revolution was important because of the way it increased human productivity and, consequently, secured human welfare. Accordingly, much philosophy of technology has established the optimistic understanding that technology, together with science, is a reliable engine of progress. However, technology does not operate as a transparent tool that we human beings can dominate and freely use for our selected purposes. Technology enters into human life; it enters the human being; it transforms society. Technological rationality formats production processes, management approaches, and economic strategies. To a critical philosophy of mathematics, it is important to consider how the implementation of a mathematics-mechanical worldview might serve particular political and economic priorities, and how this view shapes our living conditions.

Keywords Industrial revolution · Working class · Philosophy of technology · Technological optimism · Technological pessimism · Technological determinism · Technological rationality · Mathematics-mechanical worldview · Descriptive geometry · Projective geometry

The industrial revolution is sometimes described as not one, but three events. The first industrial revolution is estimated to date from around 1760 to around 1830; it is characterised by the invention of new production machineries and the construction of factories. The second industrial revolution is thought to have started around 1870, when new ways of organising and managing the production process itself were invented. The introduction of the production line is an example. The third industrial revolution is quoted as to have begun in the 1970s. It is characterised by the invention and incorporation of digital technology into the production process, as well as into all aspects of our social lives.

In this chapter, I concentrate on the first industrial revolution, which I will here on refer to as the Industrial Revolution. It is often described as having both a beginning and an end, but this appears to be a gross simplification. The Industrial Revolution took place in different time periods and with different rhythms around the world. In the Soviet Union, for example, the process of industrialisation can be related to

O. Skovsmose, *Critical Philosophy of Mathematics*, Advances in Mathematics Education, https://doi.org/10.1007/978-3-031-71375-0_3

farming that underwent huge changes during the 1930s through the collectivisation of agriculture. In China, the Industrial Revolution can be associated with the intensification of industrial production initiated by the communist leadership.

In the following, I start by looking at the Industrial Revolution as it unfolded in Britain. It was initiated by the invention of new machinery to be used in the textile industry, and soon machines were invented for all kinds of production processes. I see the Industrial Revolution as an initial step in the process of society becoming absorbed by technology. Our general living conditions have been shaped by technology, and mathematics is deeply involved in this shaping. Due to technology, we are witnessing a mathematisation of society.

3.1 An Industrial Revolution

Many inventions contributed to the Industrial Revolution taking place in Britain. Machines to make textile production more efficient were invented. In 1733, the flying shuttle was patented by John Kay (1704–1779), who came from Lancashire, in North West England. The flying shuttle made it possible to weave wider fabrics; it turned the weaving process into a much faster and more efficient process, and facilitated the further mechanisation of weaving. In 1764, James Hargreaves (1720–1778), also from Lancashire, invented the spinning jenny. It worked in a similar way to the traditional spinning wheel, but it was faster in producing the threads. A spinning frame for producing threads, constructed by means of wood and metal, was invented by Richard Arkwright (1732–1792), again from Lancashire. He patented his invention in 1769. Initially the spinning frame was operated by means of horse power, which very much reduced the costs of spinning. In 1771, in collaboration with others, Arkwright started using water power as an energy resource in a spinning factory.

In 1779, Samuel Crompton (1753–1827), born in Bolton next to Manchester, invented what was later called a spinning mule. It was a hybrid, combining Hargreaves' spinning jenny and Arkwright's spinning frame. It was a crossbreed in the same way a mule is a crossbreed of a horse and a donkey. The spinning mule was able to produce finer threads than hand-spinning, and could do so in a much faster way. In the weaving process, the purpose of the loom is to keep the longitudinal threads tight in order to facilitate the interweaving of the horizontal threads. In 1785, Edmund Cartwright (1743–1823) from Nottinghamshire patented the first mechanically driven loom; this invention was improved over time, which facilitated further mechanisation of the weaving process.

The steam engine provided the principal energy supply that really set the machinery of the industrial revolution in motion—not only the machinery in the factories, but also the network of transportation needed for industry. James Watt (1736–1819) was born in Renfrewshire in Scotland. He has often been presented as the inventor of the steam engine, but this invention took place in steps. The idea of using steam as an energy resource was formulated already in Antiquity by Heron from Alexandria. In

1551, a preliminary kind of steam engine was described by de Taqi ad-Din (1526–1585) working in Cairo and Istanbul. By means of his device, it was possible to automatically rotate a spit when, for instance, roasting meat. A similar device was described in 1620 by Giovanni Branca (1571–145) from Italy. However, Branca pointed out that the device had many more possible applications: it might be used for powering pestles, pumping water, and sawing wood. In 1720, Jakob Leupold (1674–1727) from Germany presented a steam engine used for water-pumping. On top of such contributions, Watt made further improvements to the steam engine. One of the principal challenges he overcame was to design machinery that could provide rotary motions; this made steam engines the most important energy supply for the Industrial Revolution.

The first steam-powered vehicle was built in the late eighteenth century, but only around 1800 did the mobile steam engine become a practical proposition. The first functioning railway locomotive was built in 1802 by Richard Trevithick (1771–1833), born in Cornwall. Several improved visions of the locomotive soon followed, one being the Puffing Billy, built 1813–14 by William Hedley (1779–1843) in a small city near Newcastle upon Tyne. In 1830 the Liverpool and Manchester Railway opened; it was built by George Stephenson (1781–1848), born in Northumberland. The railway was used for transporting passengers as well as goods.

Arkwright was one of the first to bring the textile industry into factories. The first energy resources were horse power and water power, but in 1781 he opened the world's first steam-driven textile machinery in Manchester, and soon steam power was used in most factories. During the early 1800s, Manchester was nicknamed Cottonopolis because of its huge number of textile factories.

3.2 Working-Class Conditions

While production based on handcraft made it natural for craftsmen to work at home, the size of the new machinery created the need for a workplace of tremendous magnitude: a factory. In order for craftsmen to work at the factory, they needed to requalify—which, for some, meant disqualification. What was needed to operate a machine was quite different from what was needed to manufacture the items by hand. Craftsmen became workers.

The bigger machines needed power to function. The classic power resource in craftmanship was the human body. For a long time, wind power and water power had complemented human muscle power. However, essential for the Industrial Revolution was the power resource established by the steam engine. This engine needed coal, and in order to ensure easy access to coal it was important to construct the factories close to existing coal mines.

It was necessary for the workers to live close to the factories; they needed to be in walking distance. In Northern England the network of coal mines and factories set the living conditions for the working class. This "topology" of the Industrial

Revolution created a most desolate place for living: smoke from the chimneys covered the sky, while black dust covered the ground. This topology facilitated economic exploitation, which dumped the new working class into extreme poverty.

Living conditions for workers and their families were horrific. Friedrich Engels (1820–1895) was the son of a wealthy German family living in Barmen, close to Düsseldorf. His father was the co-owner of big textile factories in both Barmen and Manchester. In 1842, Engels was sent to Cottonopolis to work in the office of one of the family's factories. There was much to learn for Engels, but he learned something quite different from that which his father had expected.

In Manchester, Engels met Mary Burns, who worked in one of the Engels' factories. They became close friends, and she took Engels around the working-class districts of Manchester. Engels was shocked by what he saw: the poverty, the extensive use of child labour, the many serious accidents taking place at the workplaces, and the brutal economic exploitation of workers. Engels wrote about his experiences, which were published in the *Deutsch–Französische Jahrbücher*, edited by Karl Marx. One of Engels' articles had the title "*Umrisse zu einer Kritik der Nationalökonomie*" ("Outline of a Critique of National Economy"). It might have been this article that turned Marx definitively away from the left Hegelian position that saw a critique of religious and political preconceptions as a source for social changes. Instead, Marx, as pointed out by Engels, recognised the change of economic structures as the driving force in social development.

Engels' texts were brought together in the book *Die Lage der arbeitenden Klasse in England* (*The Condition of the Working Class in England*), which was written in 1844 after Engels had returned to Barmen. The book was published in 1845. Here, Engels provided a systematic account of the living conditions of workers and their families. The account was based on Engels' own observations as well as on official documents, reports, and statistics.

Engels (1999, p. 34) started his account by making some observations with respect to big cities like London, Edinburgh, and Glasgow before concentrating on Manchester. He pointed out that, when trying to accelerate industrialisation, a big city has important advantages. It provides roads, railroads, canals, and a labour force; it offers a market to which the buyers are flocking, and ensures a direct communication line with those who supply the raw materials. Engels pointed out that a city like Manchester has a peculiar construction by sharply separating the working-class quarters from the part of the city reserved for the middle class—meaning that a person may live for years in Manchester "without coming into contact with a working-people's quarter or even with workers" (p. 57). Middle-class people might not have any notion of the living conditions of working-class people. Engels' colleagues in the factory office might be unaware of the living conditions of the workers they employ. But Engels had taken the time to visit those parts of the city where the workers lived, and he came to know the poor areas in Manchester better than he knew his own city of Barmen.

Engels observed that for working-class people in general the food was bad, and in many cases "almost unfit for use". It was insufficient in quantity, and in extreme cases "death by starvation results" (p. 85). The living conditions of the workers were also poor, with cheap housing, cheap clothes, and little or no access to doctors. According to Engels, the working-class in great cities were submitted to a graduate scale of conditions in life: in the worst cases reaching even "homelessness and death by starvation" (pp. 85–86). Engels compared the workers' conditions with that of being a slave, and concluded that the bourgeoisie were "far better off under the present arrangement than under the old slave system" (p. 92). One reason was that now the bourgeoisie could dismiss its employees at whatever time without sacrificing invested capital. Contrary to this, a slave is bought and represents a more permanent investment.

Engels was shocked by the huge number of industrial accidents, many of them fatal, that occurred at workplaces. Safety measures in the factories were far from adequate. However, accidents were only one of the many causes for the reduced life expectancy of people from the working-class. Another cause was long working hours, which exhausted the workers beyond their limits. A normal working day was from 12 to 16 hours long. Low salaries also shortened their life expectancy directly, as it was hardly a sufficient amount to secure adequate quality or amounts of food. Starvation in the world's most advanced industrial country was a real problem for the population who created the wealth.

Some of the new machinery made it possible for a person with less physical strength than was previously necessary to be used in the production process. Women and children could be exploited, and Engels reported that women were often forced to return to the mill after just three or four days of confinement. During dinner time "they must hurry home to feed the child and eat something" (p. 153). The deplorable conditions for women had a direct impact on the health of their children. With reference to the *Report on the Commission into the State of Large Towns and Populations District* from 1844, Engels pointed out that in Manchester "more than 57% of the children of the working class perish before the fifth year" (p. 118). The exploitation of children as a workforce was extreme. It did not allow them to experience childhood, and it made going to school impossible. When reaching nine years of age, children would have to work six and a half hours per day; when reaching their thirteenth year, they were required to work twelve hours per day (see p. 160). Engels added that, in Manchester, even children from five to six years old were employed at mills.

Engels accused English society of murder, and he stated: "When one individual inflicts bodily injury upon another, such injury that death results, we call the deed manslaughter; what the assailant knew in advance that the injury would be fatal, we call his deed murder" (p. 107). Then he continued:

> [W]hen society places hundreds of proletarians in such a position that they inevitably meet a too early unnatural death, one which is quite as much a death by violence as that by the sword or bullet; when it deprives thousands of the necessaries of life, places them under conditions in which they *cannot* live [...] knows that these thousands of victims must perish, and yet permits these conditions to remain, its deed is murder just as surely as the deed of the single individual... (p. 107)

Engels found that society knew the impact of what it was doing: the knowledge was expressed, although also somehow hidden, in official reports and statistics. By publishing *The Condition of the Working Class in England* Engels wanted to make it impossible to ignore the existence of the murder that was taking place by shortening the life expectation of exploited people. We were dealing with murder even though the death of a person from the working class did not seem what it is, "because no man sees the murderer, because the death of the victim seems a natural one, since the offence is more one of omission that of commission. But murder it remains" (pp. 106–107).

Karl Marx (1818–1883) also made sharp observations with respect to the working conditions created along with the Industrial Revolution, and some of his observations concerned the very content of the working process. In *Capital: A Critique of the Political Economy,*[1] first published in German in 1867, Marx (1990) wrote: "In handcrafts and manufacture, the worker makes use of a tool; in the factory, the machine makes use of him" (p. 548). In handcraft the workers move the instruments, while in the factory it is the "movements of the machine" that the worker must follow (p. 548). And Marx added:

> Factory work exhausts the nervous system to the uttermost; at the same time, it does away with the many-sided play of the muscles, and confiscates every atom for freedom, both in bodily and in intellectual activity. Even the lightening of the labour becomes an instrument for torture, since the machine does not free the worker from the work, but rather deprives the work itself of all content. (p. 548)

According to Marx, the factory physically oppresses workers, and simultaneously operates with economic cruelty. In *Capital*, Marx pointed out that capital did not emerge because one merchant did better business than other merchants. Capital emerged due to the fact that, through their work, the workers added more value to the items they produced than they received in payment for their work. The workers in the coal mines added value to the coal by digging it up so that it could be sold. The workers in the textile factories added value to the textile by turning textiles into cloth. In no such cases did their salaries add up to the value that they added to the product. This "surplus-value" was the source of the capital. Capital only emerged through exploitation of the workers.

3.3 Philosophy of Technology

Critical reflections about the socio-economic role of machines and technology were formulated by Engels and Marx. However, it was only in 1877 when Ernst Kapp published the original German version of *Elements of a Philosophy of Technology* (*Grundlinien einer Philosophie der Technik*) that a philosophy of technology got a distinct formulation, and Kapp is often considered the first philosopher of technology. Kapp (2018) used the word *Technik*, which could be translated as "technique", but soon "technology" became a more common expression.

[1] See Marx (1990, 1991, 1992).

Kapp came from Preussen, but after he published a controversial article in 1849 he emigrated to the USA, where he settled in a German community in Texas. Kapp was a self-declared freethinker, and in Texas he was elected president of an association of German freethinkers: *Die Freie Verein* (*The Free Society*). The association had as some of its principles that everyone should be paid equally for equal work, that slavery should be abolished, and that state and religion should be kept strictly separated.

Kapp saw technology as a way of compensating for limitations of the human body. He considered technology to be an extension of the human body, and he used the expression "organ projections" for such extensions. A hammer was seen as an extension of the human fist: by means of a hammer made of a stone, one could do more powerful hammering than with one's fist. The cupped hand—used, for instance, when drinking water from a river—was projected into a bowl. An axe was seen as an extension of the human arm. It added tremendous force to the arm: it could be used for cutting down a tree, for killing an animal, or for killing an enemy. Teeth and nails were seen as projected into knives.

Kapp also interpreted a bridge as being an extension of the human body. When walking, one might get to a small spring that one might be able to cross by taking a big step. One might also get to a river where one needs a bridge for crossing, and Kapp saw bridges as projections of the human hip. Kapp paid particular attention to the telegraph, and he saw the wire as an extension of the nervous system. At that time, the railway—and in particular, the locomotive—was considered a principal example of technology. A locomotive does not look like any of the human organs, but it can be seen as an extension of muscular power. One can speculate about how Kapp would interpret the radio (maybe as an extension of the human voice) and the aeroplane (maybe as an extension of the human jump). If we consider the computer, it seems rather natural to look at it as an extension of the human brain.

The idea of organ projections conceptualises technology as something we human beings are in control of. In the same way as we are in control of the fist and the arm, we are in control of a hammer and an axe. Technology is seen as a huge toolbox that we can make use of in our struggle against nature, which otherwise might appear threatening and hostile. Technology is needed for human beings to dominate nature and to make use of it. Kapp's metaphorical interpretation of technology as organ projection established an optimistic understanding of technology: to create extensions of the human body appeared to simply be a good thing to do. Kapp's philosophy of technology, associated with the technological optimism, was implanted in the modern project. The fact that technology might develop into a serious threat to human beings was not part of Kapp's philosophy of technology, nor of the modern outlook.

In *The Technological Society*, first published in French in 1954, Jacques Ellul (1964) described technology as an all-embracing system that encapsulates human life. Technology has a development of its own; it is not under human control, but rather it controls human life. Ellul pointed out that, with the word "technology", he did not refer just to tools and machines. To him, technology also involved a broad set of structures and procedures. Ellul pointed out that technology has developed into new forms without any resemblance to previous forms of technology. He did not see technology as a passive and obedient human instrument. Technology had a life of its own. It is autonomous. It is active. It sets conditions for human life.

According to Ellul, technology absorbs human beings and transforms society. From being an external instrument, technology becomes an acting and autonomous subject. We live in a technological world, which develops according to its own logic. Human beings get overpowered by technology and become its objects. Ellul synthesised his observations by stating that "our civilisation is constructed *by* techniques [...], *for* techniques [...] and *is* exclusively technique" (pp. 127–128).

In *The Technological Society*, Ellul weaved technological determinism together with pessimism. Technological development does not imply that we, human beings, gain the freedom to do what we want. Instead, we are losing freedom. While we may free ourselves from urgent material scarcity, Ellul warned that this victory may have been gained "at the price of an even greater subjection to the forces of artificial necessity of the technical society?" (pp. 428–429). The technological society operates according to its own laws; it sets its own agenda; and it establishes necessities that determine our lives.

Other philosophers than Ellul have described technology as all-embracing social structuring. In the *One-Dimensional Man*, Herbert Marcuse (1964) presented a gloomy picture of the actual functioning of sciences and technology. The output becomes a one-dimensional rationality, which he also referred to as a technological rationality. This rationality is manifested in the concept of science advocated by logical positivism.[2] One-dimensional thinking focuses on the positives given. It ignores alternatives to what is taking place, and in doing so it becomes non-critical. There is no transgression to be found in positive thinking; it leaves established realities unchallenged.

Marcuse disagreed completely with approaches in analytic philosophy which were preoccupied by interpreting everyday uses of language. He could only smile at the misplaced philosophic energy put into the analysis of a statement like "the broom is in the corner". The analytic investigation of this statement occupied Ludwig Wittgenstein (1953), and according to Marcuse much analytic philosophy continued to preoccupy itself with similar linguistic trivialities.[3] Wittgenstein had emphasised that philosophy should leave everything as it is,[4] but according to Marcuse, this causes philosophic self-destruction. Marcuse reminded us that the reality of today (meaning the reality of the early 1960s) is still that "of the gas chambers and concentration camps, of Hiroshima and Nagasaki, of American Cadillacs and German Mercedes, of the Pentagon and Kremlin, of the nuclear cities and the Chinese communes, of Cuba, of brainwashing and massacres" (p. 180). That analytical philosophy becomes one-dimensional is not only ridiculous, it is also tragic.

---

[2] See Sect. 2.5 for a discussion of the scientific worldview associated to logical positivism.

[3] See Wittgenstein (1953, §60).

[4] See Wittgenstein (1953, §124).

Marcuse found that critical thinking needs to be rooted in the metaphysical dimension of language—the dimension that was abandoned by positivism. Metaphysical thinking includes transgressive potentials. According to Marcuse, such potentials might be found in art, and should philosophy become critical, such potentials need to be found there as well. Critical thinking is two-dimensional.

Marcuse found that, due to technology, we have created material affluence, which ensures freedom from immediate needs. This tends, however, to maintain one-dimensional thinking, and to form one-dimensional human beings. Western countries have been claimed to be democratic; they are supposedly places where people can enjoy their freedom. But according to Marcuse, this is an illusion. The freedom in a capitalist society is a freedom kept on a short leash. As consumers, we can choose between several items if we can afford them; as workers we can try to qualify for different jobs, but we have to submit ourselves to the labour market; and as citizens we can choose where to cast our votes, but the available political alternatives are located within a limited spectrum. Such limitations of social life remain hidden beneath the surface of growing material welfare. Technological improvements annihilate the search for socio-political alternatives and enhance non-critical thinking.

One-dimensional thinking turns reality into an unquestionable given. This brings about totalitarianism, although not of a form based on external control and physical violence. In highly technological societies, the totalitarianism formed by one-dimensional thinking is much more sophisticated; it is acted out by making human beings one-dimensional. The technological rationality, operative for the domination of nature, has turned out to be just as operative for the domination of human beings.

With the help of Ellul and Marcuse, we have reached an interpretation of *technology as an all-embracing structuration of human life and human beings*.[5] This interpretation is radically different from the interpretation of technology as a tool or as organ projection. I will further characterise the interpretation of technology as an all-embracing structuration with the following four comments:

First, I use the notion of technology to refer to material or mechanical structures, to organisational procedures, as well as to management approaches. The Industrial Revolution showed the many features of technology: new machines were constructed, new working conditions were established, new forms of control and exploitation became possible, and new profit-generating principles was implemented. We are dealing with an all-embracing configuration of life conditions for workers and their families, and for capital owners as well.

Second, technology seems to gloss over social conflicts and contradictions. Political oppression and economic exploitation of larger social groups of people can be overlooked, even by the groups themselves, due to the minor improvements in general life conditions: electricity is installed, cars become available, new means

[5]Giddens (1984) presents structuration as a crucial notion for interpreting social phenomena. However, I do not assume Gidden's interpretation of the notion.

of communication become accessible, news and entertainment programmes are broadcast, and healthcare programmes are improved. Technology improves living conditions, even for the groups of people that continue to be exploited, also by that very technology. Technology comes to serve the preservation of the capitalist order of things. It comes to function as a political pacifier.

Third, that technology tends to ensure general improvements in welfare might appear to be true; however, it is only so if we consider what is taking place in richer communities in the world. In poorer communities, we might witness cases of oppression and exploitation that reach new extremes precisely due to technological development. Technology continues to ensure welfare for some, while impoverishing others.

Fourth, technology not only forms our material living conditions, it also forms our thoughts and actions. Technology forms what we institutionalise as knowledge and the way we apply it. The recent development of artificial intelligence is the most obvious example of this. The diversity of applications of artificial intelligence shows how our living conditions become structured by technology.[6]

3.4 Mathematics as Technological Rationality

The construction of machines was essential for the Industrial Revolution. The first machines used in the textile industry might be described as small-scale ingenious inventions, but soon the machinery grew in size and complexity. The Industrial Revolution brought about a new demand for engineering competences, which the universities at that time were unable to supply. New types of institutions for further education with a technical profile were established. In Germany these institutions were called *Technische Hochschule*, but more recently many of them became renamed as *Technische Universität.*[7] In England, The Royal Polytechnic Institution, later to be named the Polytechnic of Central London and now called the University of Westminster, was established in 1838. Its aim was to educate engineers, and soon

[6]The philosophy of technology has evolved tremendously. Ihde (1990) related technology and life-worlds in a new form of material hermeneutics. Latour (1993) elaborated on an actor-network theory that integrates reflection on technological artefacts, discourses, and human life-conditions. Beck (1992, 1999) coined the notion of risk society, by means of which he expressed an important insight: technology is not only a means for solving crises, but also a generator of crises. Many more have contributed to the philosophy of technology; see for instance: Coeckelbergh (2020), Dreyfus (1972), Dusek (2006), Feernberg (2010), Flusser (1999), Ihde (1998), McLuhan (2001), and Stork (1977). The philosophy of technology has developed into different domains; one is the philosophy of artificial intelligence, which overlaps with a philosophy of mathematics. See, for instance Boden (1987) and Coeckelbergh (2022).

[7]The Technische Universität Carolo-Wilhelmien opened in 1745, Technische Universität Berlin in 1770, Technische Universitäy Dresden in 1828, Technische Universität Chemnitz in 1836; Rheinische-Westfälische Technische Hochschule in 1870, and Technische Universität Darmstand in 1877. In this listing I have used the actual names of the institutions.

after its opening, similar institutions opened in some of the larger cities around England. In 1828, a group of business people and industrialists in Manchester met to take the first steps to open an institute concentrated on mechanics. This initiative led to the formation of the University of Manchester's Institute of Science and Technology, now incorporated into the University of Manchester. In France the École Polytechnique had opened in 1794; in 1804 it was militarised by Napoleon, and the aim of the school became to train civil as well as military engineers.

During the scientific revolution, mathematics was crucial for establishing a mathematics-mechanical worldview. This worldview was likewise important for all kinds of engineering construction. Calculus was soon recognised as being crucial for any kind of engineering, and the establishment of engineering education created a new demand for a coherent presentation of calculus.[8] For both the scientific and the Industrial Revolution, the mathematics-mechanical worldview was essential. Mathematics was indispensable for grasping how nature was operating, as well as for completing technological design. Mathematics became a key part of the technological rationality.

One of the means for carrying out technical design is descriptive geometry, also referred to as projective geometry. The construction of the impressive architecture in Antiquity in India, China, Japan, Greece, Rome, Africa, South America must have drawn on elaborated competence related to projective geometry.[9] However, it was the Italian architect Filippo Brunelleschi (1404–1472) who more systematically studied perspective drawing, an important feature of projective geometry. The main purpose of this geometry is to clarify the relationship between a three-dimensional object to be constructed and a two-dimensional representation of this object, the blueprint for the construction.[10] Contrary to Euclidean geometry, projective geometry operates with an infinitely faraway point, a vanishing point, where parallel lines do meet. Projective geometry treats the vanishing point as a real geometric point similar to any other points, meaning that it makes sense to state that parallel lines actually meet in the vanishing point. In 1765, Gasper Monge (1746–1818) was working on the planning of a military fortification, and as part of this planning he clarified how the object to be constructed can be presented in two-dimensional drawings, where the object is seen from different perspectives.

---

[8]Such a presentation was given by Cauchy (1821). See Sect. 9.2 for more comments about Cauchy's presentation of calculus.

[9]In Antiquity the relevance of mathematics for construction was clearly recognised by the Roman architect and engineer Vitruvius living in the first century before Christ. In the book *De Architecture* he addressed the construction of houses, aqueducts, mills, pumps, and central heating (see Vitruvius, 1960).

[10]An important contribution to the development of projective geometry was made by Desargues (1591–1661), who demonstrated an important theorem, now known as Desargues' theorem. In Antiquity one finds observations related to projective geometry made by Pappus from Alexandria.

This is the reason that Monge is often considered the inventor of descriptive geometry.[11]

The Industrial Revolution is not only characterised by the construction of the new production machinery, but is also associated with the formation of the capitalist order of things. This order is characterised by ways of dealing with economic transactions. A new form of technological rationality was brought into operation. This rationality materialised as "business machines".

IBM is an abbreviation for International Business Machines. When business reaches a certain level of complexity, it cannot be properly handled by paper-and-pencil calculations; automatic procedures and techniques need to be implemented. IBM was founded in 1911 as the Computing-Tabulating-Recording Company, but in 1924 it changed its name to International Business Machines. IBM became a manufacturer of tabulating machines, which were designed to handle information represented by punched cards by a combination of holes. IBM also developed the electric typewriter, the electromechanical calculator, and later the personal computer. A business machine can be seen as the embodiment of technological rationality.

Science, as well as technology, has been presented as engines of progress. Consequently, mathematics could be glorified as embodying a progressive technological rationality. Mathematics appears to be a success!

However, I find it extremely dangerous to associate technological rationality with technological optimism. It is a rationality that operates in the middle of economic and political conflicts and contradictions. This rationality makes it possible to dominate nature; to extract natural resources for securing human welfare; and to treat illnesses in a systematic way. It is this very same rationality that makes it possible to dominate other groups of people; to exploit both human beings and nature beyond their limits; and to make new war technologies possible. It is this rationality that made the Holocaust possible. Technological rationality is involved in all kinds of social complexities. It needs to be critically addressed, and so does mathematics.

References

Beck, U. (1992). *Risk society: Towards a new modernity*. Sage Publications.

Beck, U. (1999). *World risk society*. Polity Press.

Boden, M. (1987). *Artificial intelligence and the natural man* (2nd ed.). MIT Press and Basic Books.

Cauchy, A.-L. (1821). *Cours d'Analyse*. De L'Imprimerie Royale.

Coeckelbergh, M. (2020). *Introduction to philosophy of technology*. Oxford University Press.

Coeckelbergh, M. (2022). *The political philosophy of AI*. Polity Press.

[11] Whitehead (1907) presented an axiomatisation of projective geometry, which in this way turned into a regular type of geometry in line as Euclidian and non-Euclidian geometries.

Dreyfus, H. (1972). *What computers can't do: The limits of artificial intelligence*. MIT Press.

Dusek, V. (2006). *Philosophy of technology: An introduction*. Blackwell Publishing.

Ellul, J. (1964). *The technological society*. Vintage Books.

Engels, F. (1999). *The condition of the working class in England* (Edited with an introduction and notes by David Mclellan). Oxford University Press.

Feernberg, A. (2010). *Between reason and experience: Essays on technology and modernity*. MIT Press.

Flusser, V. (1999). *The shape of things: A philosophy of design*. Reaktion Books.

Giddens, A. (1984). *The constitution of society: Outline of the theory of structuration*. University of California Press.

Ihde, J. (1990). *Technology and lifeworld: From garden to earth*. Indiana University Press.

Ihde, J. (1998). *Philosophy technology: An introduction*. Paragon House.

Kapp, E. (2018). *Elements of a philosophy of technology* (Edited by Jeffrey West Kirkewood and Leif Weatherby, translated by Lauren K. Wolfe, afterword by Siegfried Zielinski). University of Minnesota Press.

Latour, B. (1993). *We have never been modern*. Harvard University Press.

Marcuse, H. (1964). *One-dimensional man*. Beacon Press.

Marx, K. (1990). *Capital: A critique of political economy, Volume I* (Introduced by Ernest Mandel. Translated by Ben Fowkes). Penguin Books.

Marx, K. (1991). *Capital: A critique of political economy, Volume III* (Introduced by Ernest Mandel. Translated by David Fernbach). Penguin Books.

Marx, K. (1992). *Capital*: A *critique of political economy, Volume II* (Introduced by Ernest Mandel. Translated by David Fernbach). Penguin Books.

McLuhan, M. (2001). *Understanding media: The extension of man*. Routledge.

Stork, H. (1977). *Einführung in die Philosophic der Technik*. Wissenschaftliche Buchgesellschaft.

Vitruvius. (1960). *The ten books on architecture* (M. H. Morgan, Trans.). Harvard University Press.

Whitehead, A. (1907). *The axioms of descriptive geometry*. Cambridge University Press.

Wittgenstein, L. (1953). *Philosophical investigations*. Blackwell Publishers.

Chapter 4
Globalised Exploitation

Abstract The so-called "big discoveries" started a race to establish colonies worldwide. Colonisation harboured imperialism, which facilitated brutal exploitation through the extraction of natural and human resources. Attempts to "justify" colonisation gave rise to further racist explanations. With the phrase "the irony of globalisation", I refer to some ambiguity encapsulated in this notion. On the one hand, globalisation seems to indicate solidarity, to embrace the perception that we are all in it together, and to signal a universal concern for each other. On the other hand, processes of globalisation demonstrate brutal forms of exclusion and exploitation. The globalised world is a world apart. Processes of globalisation include the automatisation of production processes, financial transactions, and the free movements of ownership. Such processes are rooted in mathematics-based technologies. Mathematics is deeply embedded in global networking and in all its accompanying ambiguities.

Keywords Globalisation · Discoveries · Colonisation · Racism · Irony of globalisation · Forth world · We-them dichotomy · Genocide · Exclusion · Exploitation · Automatisation · Global networking

Globalisation has been claimed to bring the world together. What happens in one part of the world will immediately be communicated to other parts; through globalisation the remote corners of the world will disappear; we are all getting closer together.

The concept of the "global city" has been applied to New York, London, and Tokyo. The huge metropolis can be seen as an international phenomenon, rather than a national city. They encompass tremendous economies, culture, politics, religion, and ethnic diversities.[1] The notion of globalisation has also been captured by the notion of the "global village", which connotates nearness and familiarity: we know about each other; we share our destinies; and we are connected by solidarity.[2]

---

[1] See Sassen (1991).

[2] The term "global village" was coined by McLuhan (1962).

O. Skovsmose, *Critical Philosophy of Mathematics*, Advances in Mathematics Education, https://doi.org/10.1007/978-3-031-71375-0_4

The expression "global village", however, conceals other faces of globalisation. Globalisation is far from an expression of broad solidarity; it encompasses brutal forms of exploitation; it generates mobility and turns the labour-maker into an international supermarket; it adds to the translocation of people in the name of "progress"; and it leads to obstruse hostile treatments of immigrants and refugees.

In order to better observe these different faces of globalisation, I am going to take a look at the so-called "big discoveries" that have been interpreted as first steps into a globalised world. The narratives about the "big discoveries" are incorporated into the legend of Modernity being a manifestation of universal progress. This legend I want to deconstruct by pointing out what I refer to as the irony of globalisation: the globalised world is a world apart.

I conclude the chapter by considering to what extent mathematics is involved in the processes of globalisation and how it might contribute to the irony of globalisation.

4.1 Discoveries

According to the Eurocentric narrative, Modernity was anticipated by the Renaissance. It was a time of discovery and exploitation. Not only were discoveries directed towards remote areas of the globe, they were also directed towards the human body itself. In architecture, art, music, and literature new forms of human expression were explored.

In *On the Fabric of the Human Body*, Andreas Vesalius (1514–1564) presented detailed anatomical descriptions of the human body (see Vesalius, 2007). Explorations of the human body were also conducted in painting and sculpture, such as the work of Leonardo da Vinci (1452–1519) and Michelangelo (1475–1564). In 1595 an anatomic theatre was opened in Padua, and public dissections became popular. In architecture new impressive constructions were created, St. Peter's Basilica in Rome being an outstanding (and extremely expensive) example. Its construction commenced in 1506 and was only concluded in 1615. In literature one finds new forms of expression, such as the work of Niccolò Machiavelli (1469–1527), Erasmus of Rotterdam (1466–1536), and William Shakespeare (1564–1616). Music developed in new formats by rejecting some of the constraints imposed by the Medieval tradition that music should serve the glorification of God. Although printing had already been invented in China, it was reinvented in Europe, and in 1450 the first Bible was printed by Johannes Guttenberg. Printing helped enormously to propagate the new ideas and insights that shaped the Renaissance.

The story about the "big discoveries" has largely been told—and believed—from a European perspective, and let me start to summarise some of its chapters. In August 1492, the Italian explorer Cristoforo Colombo (1451–1506), who became anglicised as Christopher Columbus, set out from Palos de la Frontera, a town located in Andalusia in Spain. His expedition was supported by the Spanish crown, and the hope was to find a new route to India by going west. Spain had good reasons for searching out such a route. First, European trade with India was growing and it was lucrative. Italian merchants, many of them from Venice, benefited from this trade using the eastern route to India. This route was well-established, particularly by the Arabian traders. Spain, however, had difficulty in benefiting from this easterly connection. Opening up a western route to India, although much longer, made sense since valuable goods such as precious stones and spices did not take up much space. In this case the distance of transportation was not such an important parameter. It was known by then that the earth was round, although this was still not generally accepted as common knowledge. Among the sailors on Columbus' ships there might well be some who worried about sailing too close to the edge of the earth. It was in fact necessary for Columbus to do some falsifications of the calculations of his positions to make it less western than it really was.

Columbus' expedition did not reach the edge of the earth, but on 12th of October 1492 he saw land. He assumed that it was an Indian island, and consequently the indigenous people they met, he called Indians. The land he had reached was, however, what we now know as the Bahamas. Assuming that he had completed the mission and found the western route to India, he returned to Spain in March 1493, where he was welcomed as a hero. One could imagine that Columbus might have had some self-doubt as to whether his mission did in fact discover the western route. He might well have had a good estimate of the size of the earth, as well as of the distance he had been sailing westwards. Naturally he could not know how far India might be off towards the east, but the distance he had crossed to reach India must, also to him, have seemed surprisingly short. Columbus made more voyages to his "India". On one voyage he explored the islands of Cuba and Jamaica; another voyage brought him to the South American continent, and finally he explored the coastline of Central America.

The Portuguese had explored the west coast of Africa and in 1481 a Portuguese ship had sailed up the Congo River. The Portuguese exploration towards the south continued, and a few years later the Cape of Good Hope was reached. In July 1497, Vasco da Gama (1460–1524) left Lisbon for a voyage that would bring him to India, and he arrived there in May 1498. The Portuguese continued to explore possible routes to India, and Pedro Álvares Cabrall (1467/1468–1520) had taken a much more western route to round Africa than usual. On April 1500, he saw land; it was the coast of Brazil.

The Italian explorer Giovanni Caboto (1450–1500), known as John Cabot, set out from Bristol, and in June 1497 he reached the cost of North America. It is not precisely clear exactly where he landed; locations such as Newfoundland, Labrador, and Maine have been suggested. On his return to Bristol, Cabot travelled to London to report to King Henry VII the result of his expedition, and Cabot was awarded £10 by the King for "discovering a new island off the coast of China".[3] In Bristol one can

[3] See https://www.rmg.co.uk/stories/topics/john-cabot

see a replica of the relatively small ship, named Matthew, that brought Cabot to North America and back to Bristol again.

James Cook (1728–1779) was nominated to command a scientific voyage to the Pacific Ocean, more particularly to Tahiti. One aim of this voyage was scientific, namely to observe a unique astronomic phenomenon that was predicted to take place on 3 June 1769 when the planet Venus would pass across the face of the sun. The next time the same phenomenon would occur would be in 1874. A clear sky was necessary for making the observation, and Tahiti was considered a place where the sun was always shining. Precise observations of this phenomenon would make it possible to calculate more accurately the distance between the earth and the sun. In August 1768 the expedition departed from England, and it arrived in Tahiti in April 1769. The expedition involved a number of scientists, who brought with them a collection of the most advanced instruments available at the time for making astronomic observations. After the observations were completed, Cook engaged in the second aim of the expedition, namely to search for a continent that was assumed to exist on the southern hemisphere. The assumption was based on a widely accepted, yet false, scientific idea that such a continent must exist in order to balance out the land mass in the northern hemisphere, so facilitating the regular rotation of the earth. In spite of the false scientific thinking, Cook located such a continent, named Terra Australis (southern land), which we now call Australia.

Mathematics was an integral part of the processes of discovery. Navigation is a complex activity, as is celestial navigation. Naturally the mathematical component of celestial navigation developed in stages. Such navigation was used long before the big discoveries were made; the Vikings, for instance, mastered the technique of following a definite latitude on a sea voyage. Long before they were "discovered", the people who populated the islands in the Pacific Ocean had mastered advanced forms of navigation.[4] For the Europeans, their discoveries were facilitated by the invention of new technologies such as the compass (already invented in China before Christ), the sextant (1731), and an accurate way of measuring time. Calculating the latitude of a position at sea had been possible for some time, but calculating the longitude had been a long-standing intractable problem, until in 1759 the English carpenter and watchmaker John Harrison developed an accurate marine chronometer and was eventually awarded generous compensation by the British Parliament 1714 Longitude Act.[5]

The mapping of the earth is a mathematical task. At the time of Columbus, the world map presented by Claudius Ptolemy (90–168 AD) was still the authority. In his book, *Geography*, written around 150, Ptolemy (2011), who lived in Alexandria, assumed that the earth was a sphere, and he made a careful presentation of the

[4]For an ethnomathematical study of voyaging in the Pacific, see Furuto and Monkoski-Takamure (2023). They point out: "Ethnomathematics is embedded in the people of the Pacific and in the ways we navigate the world around us. It is how we know who we are, where we've come from, and where we're going." (p. 242).

[5]See Betts (2023).

mathematical principles he used for drawing a plain map of a part of the sphere. Ptolemy was far from the first to argue that the earth was a sphere. Already in the fifth century BC, Greek philosophers had presented this possibility, as subsequently did Aristotle. Ptolemy introduced the use of longitudinal and latitudinal lines to specify a location on earth, and *Geography* includes a long list of cities and locations, specifying their longitudes and latitudes. *Geography* also presented a world map showing Europe, the Mediterranean, the top of Africa, and much of Asia. What might be located to the west of Europe and the east so Asia was not indicated. Ptolemy had significantly underestimated the distance from the West coast of Europe to the East coast of Asia. Columbus was familiar with Ptolemy's estimation, and this could be one of the reasons that he expected that he had reached India.

All the crossings of the oceans provided important contributions to the improvements of world maps. Mathematics-based land surveying turned the discoveries of the word into a mapping of the world.

The claim that the western powers were "discovering" unknown lands is an expression of Eurocentrism. What they "discovered" had existed for a long time, but the European "We" claimed to have discovered "Them". A We-Them dichotomy has always been in operation; for instance, in the Antique Greek city-states: "We", the Athenians, are different from "Them", the Spartans. People from one nation might show some animosity, if not resistance, to people from another nation. The "big discoveries" gave a pathological profundity to We-Them dichotomies, opening up the way for the march of colonisation and racism.

4.2 Colonisations

The so-called big discoveries started a race to establish colonies. The race was initiated by England, France, Spain, Portugal, and the Netherlands, but even small countries like Denmark tried to join in. At first, colonisation was guided by trading rather than military interests, as a means for extracting resources such as gold, silver, precious stones, spices, sophisticated textiles, and slaves.

The Portuguese were the first to organise slave-trading in a systematic way. In 1500 the Portuguese had "discovered" Brazil, and soon it was recognised that this country provided rich opportunities for producing sugar. The only thing needed was an adequate work force. The indigenous population was not easy to dominate and there were critical voices raised towards such attempts. Some Jesuits made it clear that it was not consistent with Christianity to enslave other human beings. The cynical solution, then, was to bring slaves from Africa, because "black" people were not considered human beings—an extremely brutal expression of the We-Them dichotomy.[6]

In the book *Greater Britain*, published in 1869, Charles Dilke (1843–1911) narrated observations from a journey through the English-speaking world that took place during 1866 and 1867. He visited the USA, Australia, Ceylon, and India. Dilke was an English politician embracing liberal ideas, supporting the labour movement as well as the feminist movements of the time. He was broad-minded and well-informed, but he was also situated within the general outlook of the time. Dilke (1869) initiated his records of his travels by stating:

> The idea which in all the length of my travels has been at once my fellow and my guide [. . .] is a conception, however imperfect, of the grandeur of our race, already girding the earth, which it is destined, perhaps, eventually to overspread. (p. 4)

Dilke talked about the "grandeur of our race", a phrase which condenses the English common-sense view of the time. He saw "our race" as spreading around the world, forming a grand brotherhood of superiority. He stuffed the expression Greater Britain full of racist preconceptions.

When visiting Virginia, he had opportunities to talk with farmers and planters and to get their opinions about whom they referred to as "negros". This brought him to observe that "our common English notions of the negro and of slavery are nearer the truth than common notions often are" (p. 12). What he could hear and see did not challenge his broadly assumed English preconceptions. In 1865 slavery was abolished in the USA, one year before Dilke visited Virginia. What were we to think of the newly installed "negro freedom"? Dilke observed that this freedom could not be taken seriously: "Nothing can force the planters to treat negro freedom save from the comic side" (p. 12). When Dilke asked a planter how black people fared under their newly obtained freedom, the answer was:

> Ours don't much like it. You see, it necessitates monogamy. If I talk about the "responsibilities of freedom", Sambo says, "Dunno 'bout that; please, mass' George: me want two wife". (p. 12)

According to the planter, freedom from slavery was not universally welcomed, since this freedom would prevent "the negros" from being married to more than one wife. Dilke also observed:

> Another planter tells me, that the only change he can see in the condition of the negroes since they have been free is that formerly the supervision of the overseer forced them occasionally to be clean, whereas now nothing on earth can make them wash. (p. 12)

The conceptions expressed by the Virgina planters resonated with Dilke's views as well as with the broadly held view of the time. In 1868, Dilke was elected to the English Parliament.

[6]Maintaining this dichotomy was not simple, if the Bible was to be taken literally. If Adam and Eve were the first human beings, how could it be that human beings could be classified in different groups?

In *The Expansion of England*, published in 1883, Robert Seeley (1834–1895), a professor in modern history at the University of Cambridge, included a series of lectures about how England developed into an empire. The book became an extreme success, and sold numerous copies. His presentation was rich in information and details, but first of all he provided an overall conception of the expansion of England as a prizeworthy enterprise that forms an integral part of English history and identity. There is no hint of any questioning of the English right to establish colonies, to create an Empire, and to dominate other groups of people. Seeley (1883) was deeply inspired by Dilke, and racial superiority is the basic axiom that lays behind Seeley's whole exposition of the expansion of England.

In *Oceana: On England and Her Colonies*, published in 1886, the historian James Anthony Froude (1818–1894) claimed that English colonisation is not principally an expression of economic exploitation, but of a belief in natural superiority—a consequence of his extensive travels to South Africa, Australia, New Zealand, and the West Indies. He was enthusiastic about the whole idea of colonisation and of building up an English empire stretching across oceans, and repeatedly he glorified the English Royal Navy for making this possible. Froude (1886) saw Oceana as "the great empire of which Great Britain has the stem, and the colonies the branches" (p. 211). It is "held together by common blood, common interest, and a common pride in the great position which unity can secure" (p. 12). Oceana has "spread her race over the globe", and has "planted them where they were now flourishing" (p. 211).

Dilke, Seeley, and Froude all contributed to the glorification of British colonialism and imperialism, not seeing them as conquering of the world, but as civilising of the world. They embraced the three features of racism: seeing the differences between races as being deep-rooted; seeing some races as being inferior; and seeing superior races as having the right to dominate inferior races.[7] They formulated a worldview that legitimised imperialism.

A principal step in the move from colonialism to imperialism was taken at the Berlin Conference that took place from 1884 to 1885. Here the colonising powers divided the world between them: Africa was parcelled up, and so were the islands of the Pacific. Germany was a latecomer to the race for colonies, but it joined the so-called "scramble for Africa" and obtained some colonies there. Belgium had also entered the scramble, and got hold of the Congo. Genocides were committed against the Herero and Namaqua people living in West Africa (now Namibia), colonised by Germany. Here the indigenous population was forced into concentration camps, where they were subjected to eugenic research, which led Eugen Fischer, who directed the studies, to recommend the genocide of such "inferior races".[8]

In 1865 slavery was abandoned in the USA, but racism was in no way eliminated. The black population was brutally intimidated by the Ku Klux Klan, a terror

[7] See Sect. 1.2 for a clarification of my use of the notion of "racism".

[8] Fisher's work was studied by Hitler when he wrote *Mein Kampf*. Later Fischer became member of the Nazi party.

organisation founded in the late 1860s. In the "Deep South", public lynching took place from the time slavery was abolished, right up to the 1970s. A horrendous documentation of this is found in the lynching postcards, sold as souvenirs. They show real photos, not just illustrations, of dead black people hanging from trees, often surrounded by a mob of smiling white people, mostly men.[9]

In the Annual Register (1800) one can read that the exploration of the globe was driven by a "new and nobler passion", not the greed for gold and conquests. Voyages had been undertaken with no intention other than "perfecting the knowledge of the globe" and the "multiplication of human comforts". The noble passion was also directed towards "the most remote and barbarous tribes", and with some indignation the *Annual Register* added that these tribes were "not only ungrateful, but jealous and hostile to their disinterested benefactors" (quotations from p. 222). The *Annual Register* was first published in England in 1758, and has been published ever since. We are dealing with enlightened comments about what has taken place during the year. In the UK in 2021, the Commission on Race and Ethnic Disparities published their report about racial and ethnic disparities in the UK. In the report it is highlighted that the slave period was not only "about profit and suffering", but also about how "there is a new story about the Caribbean experience which speaks to the slave period not only being about profit and suffering but how culturally African people transformed themselves into a re-modelled African/Britain".[10] Most sadly, echoes from 1800 can still be heard in 2021.

4.3 The Irony of Globalisation

After the Second World War, Britain did not have sufficient military power left to maintain its empire. The colonies of the British Empire, as well as of other imperial powers, gradually achieved—or fought for—independence. However, to a great extent the extraction of resources from the historical colonies was able to continue, although in a new format. The previous colonised world turned into the so-called Third World. This world, "Them", was seen as being in need of help from "We", and apparently well-intentioned programmes of support were launched. However, in parallel with such programmes, the unscrupulous extraction of both natural and human resources continued unabated.

In *As Veias Abertas da América Latina* (*The Open Veins of Latin America*), Eduardo Galeano (2021) gives many examples of natural resources extracted from South America, and from Third World countries in general, and imported into the

---

[9] See: https://withoutsanctuary.org/?ref=hybridpedagogy.org

[10] *Commission on Race and Ethnic Disparities: The Report* (2021, p. 8) also known as "The Sewell Report". The report was commissioned by and presented to the Conservative government. The statement quoted became one of the most controversial statements in the report. Nine of the ten commissioners were people of colour.

USA. He reveals the tremendous economic, political, and military interests that are associated to this extraction and importation. Many minerals necessary for both military and industrial purposes are not found in sufficient quantities within the USA's borders. In order to understand the foreign policy of the USA with respect to Latin America, it is important to be aware of the high priority of securing the country's import of raw materials. For industrial and military purposes, copper and zinc are necessary. For constructing aeroplanes, aluminium is necessary, and bauxite is the principal resource for aluminium. The USA has no bauxite resources. Manganese and nickel are important minerals, and neither manganese nor nickel is found in the USA. The USA has had adequate petrol resources, but they are running out at the same time as the demand is increasing. Hence, it is crucial for the USA to secure a reliable import of petrol.

One could think of exportation and importation as two kinds of neutral business relationships. However, exportation and importation can also be seen as expressions of power. Richer countries dominate poorer countries by defining export-import relationships as brutal as those previously set by the colonial world order. Indeed, the export-import relationships between USA and Latin America involve such brutalities. They concern the organisation of the extraction of resources, the leadership of the involved companies, and the price for the resources. In general, less than 10% of the USA's market value of an imported material is paid to the location from which the material is imported (Galeano, 2021, p. 223). The salary of the workers who are actually doing the extraction are kept to an extreme minimum, and in many cases the work is carried out under slave-like working conditions. Forms of modern slavery are common at the initial stage of an export-import route.[11] Galeano summarises his observations in the following way: "The North American economy needs the Latin American minerals like the lungs need the air" (p. 192, my translation).

In the article "Three Worlds, One Planet", Alfred Sauvy (1952) divided the countries of the world into three categories. The *First World* included economically powerful countries such as the USA, Australia, Western Europe, and Japan. The *Second World* included socialist nations such as the Soviet Union, China, and Cuba. And the *Third World* referred to the countries that did not participate in the Cold War. However, soon Third World countries were not associated with political neutrality, but with poverty and misery. Most Third World countries are former European colonies, and accordingly they are considered "economically underdeveloped".

The expression "Third World country" brings about a gross over-simplification, which is pointed out by Manuel Castells (1998), when he introduces the notion of the Fourth World. The Fourth World does not follow national borders; it represents a different geography. It includes parts of the Third World located in Africa, Latin America, and Asia, but provinces of the Fourth World can be found all over the planet. They can be found in huge metropolises like New York, London, Paris, Barcelona, and Rome. In São Paulo one finds huge *favelas*. Here, many houses

[11] For a discussion of slavery as it occurs today, see Bales (2012).

might not have running water or any form of sanitary installation; some are constructed with whatever might be at hand. In the fashionable Avenida Paulista in São Paulo, one sees homes made of plastic sacks. Many people do not even have a tent for a home, but are living on the streets. Some are forced to carry around their belongings in a shopping trolley stolen from a supermarket. However, such living conditions as these are not a particularly Brazilian phenomenon; they can be found in any metropolis around the world. These areas are provinces of the Fourth World.

Processes of globalisation provoke social and economic exclusion; they include the formation and the steady growth of the Forth World. The globalised capitalist economy is not welcoming to everybody. The functioning of this economy depends on a qualified work force and willing consumers, and the large groups of people who do not fall into either of these categories are excluded. Globalised capitalism does not function as a universal welfare system, but as a selective enclosure. Business people are travelling, researchers are travelling, experts are travelling, tourists are travelling, but not everybody has the possibility of moving around. In the globalised world, large groups of people are immobilised and fixed to their allotted place. To a large extent the Fourth World is populated with immobilised people.

By using the phrase *the irony of globalisation*, I refer to an ambiguity encapsulated by the notion of a global world. On the one hand, globalisation seems to indicate solidarity, to embrace the perception that we are all together, and to signal a universal concern for each other. On the other hand, processes of globalisation demonstrate brutal and cynical forms of exclusion, exploitation, and economic discarding. Globalisation deepens the We-Them dichotomies. The globalised world is a world apart.

4.4 Mathematics as Global Networking

What could mathematics have to do with processes of colonisation and globalisation? I am not going to argue that mathematics is a cause of exploitation and exclusion, but will highlight that mathematics might be critically involved in the processes of colonisation and globalisation.

Let us have a look at what was taking place in a coffee house that opened in 1686 in London. The owner was Edward Lloyd, and from the coffee house located in Tower Street there was a good view of the River Thames. The coffee house became a gathering place for sailors, merchants, and ship-owners. While enjoying a drink they could look at ships passing by. Some of the ships would be heading for far-away colonies, while others were returning from long voyages. Ships could be caught by storms; shipwrecking was common; and considerable time could pass between when a ship left London until one could hope to see it again. When a ship returned, Lloyd's Coffee House could be filled with excitement: Which ship was it? How long had passed since it left? Where had it been? Did is look to be returning with a heavy cargo?

Some ships just never returned. To be a sailor was a dangerous occupation, and so it was to be a ship owner and to invest in sea voyages. Sailors lost their lives; owners lost their money. Seeing a ship returning back home was an extreme relief for many. A ship not returning might be a family tragedy for some and an economic catastrophe for others. Edward Lloyd supported his customers by providing shipping news and this, in turn, facilitated much discussion about risk. The idea of sharing risks out among investors gradually emerged, and the idea of marine insurance became formulated. In 1774, the Society of Lloyd's was established, and came to operate as a regular insurance company. It became possible to insure all kinds of ships and cargo. Lloyd's obtained a monopoly on insurance with respect to slave trade, and they kept the monopoly until 1807, when slave trade was abolished.[12]

The idea of insurance and the emergence of insurance companies facilitated the economic processes related to colonisation. More capital was attracted to the business of extracting values and resources from the colonies, as the risks associated with such investments became easier to manage. For organising insurance strategies and for implementing insurance practices, mathematics is essential. Mathematics soon became a crucial part of all insurance practice, and insurance mathematics has developed into a highly elaborate and advanced mathematical discipline. The mathematical discipline of actuarial science thus evolved, not out of curiosity, but of economic necessity. Naturally, it does not make sense to think of mathematics as being the cause of colonisation and exploitation, but mathematics facilitated the economic conditioning of such processes.

Mathematics contributed to the navigation techniques important for making the "big discoveries", and mapping the world was a prerequisite for imperial domination. Cartography is not simply a neutral process of describing the world; it requires many decisions to be made. One concerns the problem of representation of the curved surface of a three-dimensional sphere onto a two-dimensional flat surface. A common projection to be used is the Mercator projection, which shrinks the sizes of countries the closer they are to the equator, making them appear smaller compared to countries further away. The mathematics used here gives Europe dominance over Africa. England, for instance, appears to be the same size as Kenya, while Kenya in fact is more than four times the size of England.[13] Another feature of the politics of world mapping is the definition of the zero degrees longitude, the prime median, and the positioning of the date line. That the prime median is the one passing through Greenwich and, consequently, that the date line was positioned at "the other side of the earth", is an expression of British imperialism. This means that "the other side" has to deal with the troubles caused by a passing date line, cutting, for instance, the same day into a Sunday and a Monday. Cartography not only provides the political mapping of the earth, it could also display the location of mineral resources and other

[12] See Williams (1944) for a discussion of the relationship between early capitalism and the slave trade of the 17th and 18th centuries.

[13] Gutstein (2006) describes how the politics of world mapping can be addressed in the mathematics classroom. For a discussion of the politics of mapping, see Debarbieux and Hirt (2022).

raw materials, which is an important source of information for both colonising and exploitation.

In *Capitalism and Arithmetic: The New Math of the Fifteenth Century*, Frank Swetz (1987) outlines the importance of new arithmetic techniques for dealing with the growth in trading at the time. Since then, mathematics has become a resource for establishing techniques that make international trading more efficient and establish the capitalist order of things. *Capitalism and Arithmetic* includes a translation of *Treviso Arithmetic*. The author and the title of this book are unknown, hence it is referred to by the name of the city, Treviso, where it was printed in 1478. Treviso is a small city just north of Venice. *Treviso Arithmetic* is the first printed mathematical textbook. It includes a careful introduction to the arithmetic techniques that are crucial for accounting. Venice, including the region around it, was a centre for international trade, in particular with the Arabic world and with India. Venice had turned into the capitalist centre of the Renaissance. Compared to the Arabic-Indian number system, the Roman number system appeared inadequate to deal with large-scale accounting, and *Treviso Arithmetic* includes a careful introduction of commercial mathematics based on the Arabic-Indian number system. The book was not meant for university studies, but for business practices. At the time Venice had many business schools where young merchants could study commercial mathematics, and students came from many countries, in particular from Germany.[14]

The economic complexities of the expanding trade between the colonising powers and the colonies had to be dealt with in a thoroughly rational way, and mathematics provided the means for doing so. The intimate relationship between colonisation, business, capitalism, and mathematics was established, and ever since then mathematics and economics have grown closer and closer together.

Today, the money transfer that is necessary for a globalised economy is executed by mathematical algorithms. If one should make a follow-up study of the relationship between capitalism and mathematics, one can hardly call it *Capitalism and Arithmetic*. A more appropriate title would be *Capitalism and Algorithms*. It is simply impossible to image the functioning of today's globalised capitalism without the use of computing, which is nothing but mathematical algorithms brought into action. The two books *Capitalism and Arithmetic* and *Capitalism and Algorithms* would provide two chapters of the study of the connection between mathematics and globalised capitalism. The book *Capitalism and Algorithms* is just waiting to be written.

One important feature of globalisation concerns ownership: Who is the owner of a factory or a company? An answer to this question has been given by Albert J. Dunlap: the stakeholders are the owners.[15] In the end it is the stakeholders that make the decisions. The people who are running the daily business, as well as the workers who produce the wealth, are eliminated from the principal decision

---

[14] I am not aware of any study that clarifies the accounting systems that were applied at the time by Arabic and Indian merchants.

[15] See Dunlap and Andelman (1996) and Bauman (1998).

processes. Such decisions could, for instance, concern the location of the factory. Cost-benefit analyses might reveal that it is beneficial to reopen it in a different location. The closing down of a factory might have disastrous consequences for the neighbourhoods and communities its workers live in, but such consequences need not be taken into consideration by the stakeholders, who focus on maximising the possible profit of their investments.

The group of stockholders is ever-changing, depending on the investments made. The mobility of ownership is brought to an extreme through day trading. A day trader may start the day with a certain amount of money. During the day, he or she might buy and sell bonds continuously. The exchange rates of bonds are pulsating: they move a bit up, down a bit, and so on. Such seismographic variations make it possible to profit during a day: one needs to buy when the bonds are slightly down, and sell when they are slightly up. Day traders might be co-owners of a company for, say, ten minutes. They might not even pay attention to which companies they might be a stakeholder in. Their responsibility for the neighbourhood where the company might be located is reduced to zero.

Imperial colonisation might now belong to the past, but the ruthless extraction of resources from previous colonies continues, although now it is organised in very different ways. The results, however, are the same: some groups of people benefit a lot from global resources, while others get almost nothing. It is an expression of irony: globalisation today maintains a scheme for preserving a most unequal distribution of the planet's resources.

Mathematics contributes to this irony, as today's processes of globalisation and exploitation are rooted in world-wide mathematics-based networking.

For a critical philosophy of mathematics, it is important to show that Modernity includes deep contradictory trends, and that mathematics is involved in all kinds of socio-political affairs. It is important to consider the full social complexity within which mathematics becomes developed and brought into action. For a critical philosophy of mathematics, it is important to recognise how mathematics is involved, for the good and for the bad, in all spheres of life.

References

Annual Register. (1800). *Annual register or a view of the history, politics, and literature for the year 1800* (Printed by William Heney, Crown Court, Aldersgate Steet).

Bales, K. (2012). *Disposable people: New slavery in the global economy* (3rd ed.). University of California Press.

Bauman, Z. (1998). *Globalization: The human consequences*. Colombia University Press.

Betts, J. (2023). *John Harrison and the quest for longitude*. National Marine Museum.

Castells, M. (1998). *The information age: Economy, society and culture. Volume III, End of millennium*. Blackwell Publishers.

Debarbieux, B., & Hirt, I. (Eds.). (2022). *The politics of mapping*. Wiley.

Dilke, C. W. (1869). *Greater Britain: A record of travel in English-speaking countries during 1866 and 1867*. Lippincott & and MacMillan and.

Dunlap, A. J. with Andelman, B. (1996). *Mean business: How I save bad companies and make good companies great.* Random House.
Froude, J. A. (1886). *Oceana: Or, England and her colonies*. Longmans, Green and Co.
Furuto, L. H. L., & Monkoski-Takamure, A. (2023). Voyaging beyond the horizon: An ethnomathematics legacy in Hawaii and the Pacific. In M. Borba & D. Orey (Eds.), *Ubiratan D'Ambrosio and mathematics education* (pp. 241–252). Springer.
Galeano, E. (2021). *As veias abertas da América Latina*. Colecão L and PM Pocket.
Gutstein, E. (2006). Math, maps, and misrepresentation. In E. Gutstein & B. Peterson (Eds.), *Reading and writing the worlds with mathematics: Towards a pedagogy for social justice* (pp. 111–123). Routledge.
McLuhan, M. (1962). *The Gutenberg galaxy: The making of typographic man*. University of Toronto Press.
Ptolemy, C. (2011). *Geography* (E. L. Stevenson, Trans.). Dover Publications.
Sassen, S. (1991). *The global city: New York, London, Tokyo*. Princeton University Press.
Sauvy, A. (1952, August 14). Trois Mondes, Une Planète. *L'Observateur*.
Seeley, J. R. (1883). *The expansion of England*. MacMillan and Co..
Swetz, F. J. (1987). *Capitalism and arithmetic: The new math of the 15th century*. Open Court Publishing Company.
Vesalius, A. (2007). *On the fabric of the human body: The organs of nutrition and generation*. Norman Publishing.
Williams, E. (1944). *Capitalism and slavery*. University of North Carolina Press.

Part II
Isolation and Glorification

Chapter 5
Mathematics as Logical Tautologies

Abstract Logicism tried to show that the foundation of mathematics is in logic. The logicist programme was launched by Frege, and elaborated in technical details by Whitehead and Russell. Logicism confronts the idea that mathematical notions and theorems are grounded in empirical observations and personal experiences; logicism confronts any such form of psychologism. A key point in the logicist programme is Frege's definition of number in terms of set theoretical notions. According to logicism all mathematical concepts can be defined by logical concepts, and all mathematical theorems can be derived from logical theorems. Consequently, logicism sees mathematics as logic. Since logical theorems can be shown to be tautologies, all mathematical theorems consequently become tautologies. Logicism inspires the idea that the proper language of science is a formal mathematical language, where the meanings of concepts are defined in terms of sets, and the meanings of propositions are defined in terms of their truth values. A similar theory of meaning has found its way into mathematics education dressed up like the Modern Mathematics Movements.

Keywords Logicism · Logicist programme · Foundation of mathematics · Psychologism · Formal logic · Set theory · Tautology · Sense and reference · Absolutism · Extensional language · Intensional language · Modern Mathematics Movements · Wonderland

During the age of Modernity mathematics played a crucial role in the scientific revolution by establishing a mathematical-mechanical worldview. This view was soon applied not only to natural phenomena, but to a range of social phenomena. Mathematics played an important role in the industrial revolution by synthesising a technological rationality. Mathematics facilitated the processes of globalisation by forming structures on which global networking was built.

That mathematics contributed to the configuration of Modernity, both for the good and for the bad, sits in the blind spot of most important philosophies of mathematics which were formulated during Modernity. The key philosophies that I have in mind are logicism, formalism, and intuitionism, which I will refer to as the Three Magi. The pathological blind spot of the Three Magi is combined with a focus

O. Skovsmose, *Critical Philosophy of Mathematics*, Advances in Mathematics Education, https://doi.org/10.1007/978-3-031-71375-0_5

on some particular ontological and epistemic questions: What is the nature of mathematical entities? How do we come to know about them? What is the nature of mathematical truths? How can we ensure that mathematical truths become absolute truths? By focusing on such questions, the Three Magi contributed to a deification of mathematics.

For a critical philosophy of mathematics, it is important to engage in broader reflections with respect to the social and political roles played by mathematics. It is important to recognise the full complexity of ways that mathematics is brought into action, and how it forms our worldviews as well as our realities.

5.1 The Logicist Programme

The ambition of logicism was to provide a solid foundation upon which mathematics could be erected as an architecture of absolute truths.[1] In order to do so, it was necessary to demonstrate that by means of a few logical notions any mathematical notion could be defined, and that any mathematical proposition could be derived from a few selected self-evident logical propositions serving as axioms of the entire logico-mathematical architecture. This would ensure the absolute truth of the whole body of mathematics.[2]

In mathematics one operates with different sets of numbers such as: *N*, the set of natural numbers; *Z*, the set of whole numbers; *Q*, the set of rational numbers; *R*, the set of real numbers; and *C*, the complex numbers. The set of whole numbers can be defined in terms of natural numbers, and the set of rational numbers can be defined in terms of whole numbers. In 1872 Richard Dedekind (1831–1916) showed how the set of real numbers can be defined in terms of rational numbers (see Dedekind, 1963). The set of complex numbers can be defined in terms of real numbers. This means that the whole complexity of numbers comes to rest on the set of natural numbers.

Logicism made an important contribution to such a programme of reduction by presenting mathematical concepts and theorems as molecular expressions composed of atomic logical components.

An important step in realising the logicist programme was to systematise logic, and in the *Begriffsschrift,* first published in 1879, Gottlob Frege (1983) tried to do just that. His ambition was to list all valid forms of logical reasoning. Apparently, this would be an infinitely long list, so how was he to make this list? Frege's approach was to provide an axiomatisation of logic, and the first step was to provide an axiomatisation of propositional logic. Frege did so, and his claim was that the theorems in the axiomatic system that he presented were precisely the valid form of

[1] The following presentation of logicism draws on Ravn and Skovsmose (2019).

[2] See Carnap (1983) for an introduction to logicism. Carnap's paper was original published in German in 1931.

reasoning in propositional logic. The next step in the logicist programme was to show how mathematical notions could be defined in terms of logical notions. A decisive move in this reduction was to show how the very concept of number could be defined by logical notions. In *The Foundations of Arithmetic*, published in German in 1884, Frege (1978) established this step. His final step was to elaborate the logico-mathematics architecture in all its technical details, and in *Grundgesetze der Arithmetik I-II*, Frege (1893/1903) showed in detail how mathematics could be presented as a deductive system based upon a foundation of logic. In his elaboration of the logicist programme, Frege was fully focused on arithmetic, showing how concepts of numbers and statements about numbers could be rooted in logic. However, soon the logicist programme was generalised to showing how mathematics in general was rooted in logic.

In *Principia Mathematica* published in three volumes, Alfred Whitehead and Bertrand Russell (1910/1913a/1913b) reworked all the technical details of a logico-mathematics architecture. Frege had invented a quite particular symbolic system for presenting logic; his formulas appeared like a sort of pictograms. In *Principia Mathematica*, Whitehead and Russell adopted a different symbolism, similar to what now has become standard in formal logic. The logicist programme reached a peak with the publication of *Principia Mathematica*, which presented logic-mathematics as a shinily polished gigantic monolith.[3]

Logicism offered its reductionist programme as a reaction to the psychologist programme, the aim of which was to show how mathematical concepts and theorems were based upon empirical experiences and observations. Empiricist philosophers insisted that all kinds of knowledge were based on sense impressions. In *A System of Logic*, first published in 1843, John Stuart Mill (1806–1873) presented an empiricist interpretation of all kinds of knowledge, including mathematics. Mill (1970) argued that a mathematical statement such as $1 + 2 = 3$ is an inductive truth, based on the very many observations showing that when we unite one element with two more elements, the result is three elements. The idea of mathematical statements being necessary truths is an illusion. They are empirical truths, although based on a broad range of empirical observations. Consequently, when one searches for the foundation of mathematics, one should search for observations and experiences. In this sense psychologism claims that the foundation of mathematics is to be found in psychology.

In *The Foundations of Arithmetic*, Frege (1978) launched a ferocious attack on Mill's empirical outlook. Frege remarked ironically that if "everything in the world were nailed down" (p. 9), then we would never have come to know that $1 + 2 = 3$ is true. Furthermore, Frege observed that if the "definition of each individual number did really assert a special physical fact, then we should never be able sufficiently to admire, for his knowledge of nature, a man who calculates with nine-figure

[3]Personally, Russell was deeply concerned about ensuring the absolute truth of mathematical statements. A cartoon has made a drama of Russell's epic search for truth, see Doxiadis et al. (2009).

numbers" (p. 10). According to Frege, the concept of number is devoid of any experiential components. This claim brought him into the Platonist camp.[4]

Platonism operates with a distinction between appearance and reality.[5] While the apparent world undergoes changes, the real world remains unchanged. The real world is not abstracted from the experienced world, which is rather fluttering shadows of the unchangeable real world. In order to understand what we perceive, we need to grasp the reality behind appearance. According to Plato, the unchangeable reality is made up of ideas. We are not able to perceive this reality through our senses, which are all directed towards appearance. Reality, however, is accessible through our rationality, by means of which we are able to grasp the properties of the entities that make up the world of ideas. According to Plato, mathematical entities like points, lines, triangles, and numbers are to be found among these entities. We do not grasp the necessary truths of mathematical theorems by means of our senses, but by means of logical reasoning. The Platonist view of mathematics was shared by Frege, who found that the mathematical reality is the world of sets.

5.2 The Logicist Definition of Number

A fundamental question is: What is a number? It is crucial for the logicist programme to find an answer to this question, as this ensures that the step from logic to mathematics becomes feasible.

In *The Foundations of Arithmetic*, Frege (1978) pointed out: "Number is not abstracted from things in the way that colour, weight and hardness are, nor is it a property of things" (p. 58). This is an explicit rejection of any form of psychologism that goes searching for the roots of mathematical notions in empirical observations and personal experiences. In order to grasp Frege's definition of number, we need to look closer at Frege's terminology. He contrasted the extension of a concept with the intension of a concept. By the *extension* of the concept of "mother", he referred to the set of all mothers. By the *intension* of the concept of "mother", he referred to the content of the notion, as for instance, "a woman who has given birth to a child". With this distinction made, Frege defined a number in the following way:

> The Number which belongs to the concept *F* is the extension of the concept "equal to the concept *F*". (pp. 79–80)

This compressed formulation needs to be read a couple of times. In the German original, Frege stated that *"die Anzahl, welche dem Begriffe* F *zukommt, ist der*

[4]Despite Frege's harsh critique of psychologism, Mill's empiricist interpretation of mathematics has inspired important developments of naturalistic philosophies of mathematics. Kitcher (1983) refers explicitly to Frege, when he formulates his philosophy of mathematics. Maddy (1997) presents a coherent naturalistic conception of mathematic. For an overview, see Maddy (2005).

[5]So did Galilei, see Sect. 2.3. For a clarification of Platonism in mathematics, see Bernays (1983). See also Ernest (2023) for a critique of Platonism.

Umfang des Begriffes 'gleichzahlig dem Begriffe F'" (pp 79–80). Frege talked about *Anzahl*, which can be translated as "number" or "number of", and about *gleichzahlig*, which literally means "same number", which is translated as "equal to". So, in the definition of *Anzahl* Frege used *gleichzahlig*. This looks circular, but it is not. To establish that two sets are equal to each other (*gleichzahlig*), it is not necessary to do any counting. In order to verify that the number of seats in a cinema is equal to the number of people in the cinema, one need not count first the seats and then the people. One only needs to verify if all the seat are taken by one and only one person and that all people are sitting.

According to Frege's definition, the number 2 is the set of all possible couples. And generally: a number is a set of sets, all similar (*gleichzahlig*) to each other.

In 1903 Russell sent a letter to Frege where he highlighted that he found himself in complete agreement with all the essentials in Frege's logicist project. In particular Russell highlighted that he also rejected any version of psychologism. He added, however, that he had encountered a contradiction in the system presented by Frege which he condensed in the following question:

> Let w be the predicate: to be a predicate that cannot be predicated of itself. Can w be predicated of itself? (Russell quoted after Heijenoort, 1967, p. 124)

Frege was shocked. Russell had identified a contradiction in the very foundation of Frege's logicist system as presented in the *Grundgesetze der Arithmetik.* In his answer to Russell, Frege said: "Your discovery of the contradiction caused me the greatest surprise, and, I would almost say, consternation, since it has shaken the basis on which I intended to build arithmetic" (Frege quoted after Heijenoort, 1967, p. 127). Frege received the letter from Russell when Volume II of *Grundgesetze der Arithmetik* was on its way to the printer. In a hastily written postscript, Frege tried unsuccessfully to solve the problem, after which he published nothing more in life.

Russell's formulation was highly compressed, but let me present the paradox using a normal set-theoretical symbolism. Let W denote the set of sets that are not members of themselves. In other words: $W = \{x|x \notin x\}$. We can then ask: Is W a member of itself? If it is, the W has the property that characterises W, namely that $W \notin W$. On the other hand, if W is not a member of itself, it has the property $\neg(W \notin W)$, which means that we $W \in W$. Thus, we have $W \in W$ if and only if $W \notin W$.

The paradox is rooted in Frege's basic idea that whatever predicate P one is dealing with, one can always talk about the extension of the predicate, *Ext*(P), which refers to the set of entities to which the predicate applies: $Ext(P) = \{x|P(x)\}$. According to Frege, the formation of sets is without any form of restriction. The paradox, however, demonstrates that such a free formation of sets cannot be maintained.

After much consideration, Russell decided to introduce a theory of types.[6] According to this, a set can only be composed of elements that are defined in

[6] See Russell (1967), where he outlined a theory of types. This article was first published in 1908 in the *American Journal of Mathematics, 20*, 222–262. In *Principia Mathematica*, Chapter II is dedicated to presenting the theory of types, see Whitehead and Russell (1910, pp. 37–65).

advance. This sounds a healthy assumption, and it prevents more suspicious sets from surfacing to form part of mathematics. Still, this healthy assumption was devasting to the logicist programme, which was to provide mathematics with a foundation in logic, and not with a foundation in logic combined with some additional heuristic rules as the theory of types appearedx to be. The situation was even worse. In mathematical analysis one defines, for instance, the supremum, *Sup*(*A*), of a not-empty set *A* of real numbers in the following way:

$$Sup(A) \text{ is the smallest } r \in R, \text{such that } a \leq r \text{ for all } a \in A$$

But here one defines *Sup*(*A*), which is a real number, by referring to the whole set *R* of real numbers. In mathematics it is generally assumed that this definition makes sense; consequently, a theory of types cannot be applied generally, but needs to include space for exceptions. Nevertheless, a theory of types was assumed by Whitehead and Russell, and with this assumption they elaborated in *Principia Mathematica* the somewhat modified logicist programme in all technical details.

After pages and pages filled to the brim with symbols, one can on page 362 in *Principia Mathematica*, Volume I read that from where we have got so far, we can now prove the mathematical statement $1 + 1 = 2$. Like so many before me, it has been taken as a given that it was Russell that inserted this comment. I imagine him doing so with a smile, knowing that it would be noticed by the many people that, like me, leaf through the many pages with formalisms just to see how it all appears. At the same time this remark captures in a nice self-ironic way the idea of the whole logicist programme, namely that simple mathematical theorems are derived from logic, meaning that if one really wants to know why $1 + 1 = 2$, then the answer is to be found in all the pages coming before page 362.

5.3 Logicist Absolutism

The overall ambition of logicism was to demonstrate that mathematical truths are absolute truths. Russell was obsessed with ensuring mathematical certainties. As a child, he was presented with Euclid's *Elements* by his seven-years-older brother Frank. Bertrand was electrified. This was beautiful! But he became disappointed when he was told that he had to accept the axioms. Frank told that if he did not accept them, they could not proceed. Bertrand accepted, and the Euclidian certainties were spread out in front of him.[7] One can see the experience of mathematical certainties as liberation for the young Bertrand. His parents died when he was still a small child, and he grew up in his grandparents' luxurious home, Pembroke Lodge, located in Richmond Park in London with a magnificent view over the Thames valley. This home was lavishly furnished with "Victorian certainties", and Bertrand might

[7] See Russell (1951).

somehow be liberated from these mouldy taken-for-granted objects by recognising the existence of mathematical certainties. They were certainties to be respected.

In 1889, Giuseppe Peano (1858–1932) published *The Principles of Arithmetic, Presented by a New Method* (Peano, 1983). Peano's original version was written in Latin, which by that time had become outdated for academic writing. However, while his language may have been old-fashioned, his ideas were definitely new. He presented an axiomatisation of the theory of natural numbers, which had not been properly axiomatised before.[8] Peano published his axiomatics before Whitehead and Russell started elaborating on the logicist programme, but they could certainly not feel satisfied with Peano's approach.

In order to understand why, let us look more carefully at Peano's axiomatisation. Peano used the following five axioms:

(1) $1 \in N$
(2) $n \in N \Rightarrow S(n) \in N$
(3) $\forall n \in N : 1 \neq S(n)$
(4) $\forall n, m \in N : S(n) = S(m) \Rightarrow n = m$
(5) $\forall P : [P(1) \wedge (\forall n \in N : P(n) \Rightarrow P(S(n)))] \Rightarrow \forall n \in N : P(n)$

N refers to the set of natural numbers, *n* and *m* to natural numbers, *S* to the successor function, and *P* to a predicate about natural numbers.[9]

The axiom (1) looks completely straightforward: it states that 1 is a natural number. The axiom (2) states that if *n* is a natural number, then the successor to *n*, *S*(*n*), is also a natural number. The axiom (3) states that whatever natural number *n* we are dealing with, 1 cannot be its successor. The axiom (4) states that if the successor of *n* and the successor of *m* are identical, the *n* and *m* will be identical as well. The axiom (5) captures the idea of mathematical induction, which is not to be confused with empirical induction. Mathematical induction refers to a particular proving strategy. For proving that a property *P* applies to all natural numbers, one starts proving that it applies to the number 1. In other words, one proves that *P*(1) is valid. Then one proves that if the property applies to a number *n*, the property also applies to the number $n + 1$. In other words, one proves that $P(n) \Rightarrow P(n + 1)$ is valid. Finally, one concludes that the property applies to all natural numbers, in other words that the theorem $\forall n \in N$: $P(n)$ is valid.

We might assume that it is important to argue carefully for the validity of mathematical induction. However, this was not Peano's approach. He simply stated the principle as his axiom (5). Such an approach was not in any way acceptable to logicism. Mathematical truths had to be demonstrated to be logical truths, and not just assumed to be true by being nominated as axioms.

---

[8]Grassmann (1861) made an important effort in axiomatising number theory, however, Peano's axiomatisation became *the* axiomatisation of the natural numbers.

[9]Peano (1983) enumerated nine axioms. Through three of them he specified that the relationship "=" is symmetric, reflective, and transitive; and by one axiom he stated that if *n* is a natural number, and $n = m$, then *m* is also a natural number. Such axioms concerning properties of "=" are now normally not considered part of the axiomatisation of the natural numbers.

What, then, is a logical truth? The notion of tautology has many common-sense interpretations: a statement is tautological if it states an irrefutable truth, such as "bachelors are not married". Tautological truths are self-evident truths. One can think of the axioms of a logical axiomatic system as being tautologies. *Principia Mathematica* starts out with the following five axioms:

(1) $(p \vee p) \Rightarrow p$
(2) $q \Rightarrow (p \vee q)$
(3) $(p \vee q) \Rightarrow (q \vee p)$
(4) $(p \vee (q \vee r)) \Rightarrow (q \vee (p \vee r))$
(5) $(q \Rightarrow r) \Rightarrow ((p \vee q) \Rightarrow (p \vee r))$

The axiom (1) states that if p or p is the case, then p will be the case. The axiom (2) states that if q is the case, then p or q will be the case. The axioms (3) and (4) seem self-evident as well. The axiom (5) appears more complex, but giving it an extra reading, it becomes self-evident. In this sense the five axioms appear to be tautologies.

The approach used by Whitehead and Russell in *Principia Mathematica* is similar to that used by Euclid in the *Elements*. The start is provided by the axioms which appear so simple that their truths can be grasped immediately. However, the concept of something being self-evident or tautological is not without its problems. To complete the logicist project in a satisfactory way, it seems important to better characterise what to understand by a tautology. But no further clarification of the notion of tautology is to be found in any of Frege's texts, nor in *Principia Mathematica*.

In 1911, Ludwig Wittgenstein (1889–1951) visited Russell for the first time in Cambridge; it was during the time when Whitehead and Russell were organising *Principia Mathematica* for publication. Very soon Wittgenstein became deeply engaged in the philosophic issue raised by the logicist programme, and he concentrated his energy on the foundation of logic. What is the nature of the solidity of this foundation? By the outbreak of the First World War, Wittgenstein enrolled as a soldier in the Austrian army. He went to serve at the Eastern front, and Russell did not hear from him. However, Russell knew that Wittgenstein had been investigating the notion of tautology, and in the final part of the *Introduction to the Philosophy of Mathematics*, Russell (1919) made the following comment:

> The importance of "tautology" for a definition of mathematics was pointed out to me by my former pupil Ludwig Wittgenstein, who was working on the problem. I do not know whether he has solved it, or even whether he is alive or dead. (p. 205)

Russell wrote the *Introduction to the Philosophy of Mathematics* when he, in 1918, was incarcerated for six months in Brixton prison for circulating pacifist views, which at that time was considered a crime in Great Britain. Jingoism created the necessity of young men "serving the country". Russell knew that Wittgenstein had been concentrating on a manuscript addressing logical issues, but, as he stated, he did not know if Wittgenstein had clarified further the notion of tautology, nor even if he was alive or not.

Fig. 5.1 Truth tables for the logical connectives

p	q	$\neg p$	$p \vee q$	$p \wedge q$	$p \Rightarrow q$	$p \Leftrightarrow q$
T	T	F	T	T	T	T
T	F	F	T	F	F	F
F	T	T	T	F	T	F
F	F	T	F	F	T	T

Fig. 5.2 The truth table for axiom (2) in the *Principia Mathematica*

p	q	$p \vee q$	$q \Rightarrow (p \vee q)$
T	T	T	T
T	F	T	T
F	T	T	T
F	F	F	T

Wittgenstein survived. He had also got a firm grip of the slippery notion of tautology. During the whole war period, he took with him the manuscript that later became published as *Tractatus Logico-Philosophicus*. When the war finished, the manuscript was also concluded. Wittgenstein became a prisoner of war, and for nine months he was confined in a prisoner of war camp in Italy. In 1919 he managed to communicate with Russell; they met in the Hague, where Wittgenstein was able for the first time to show the manuscript to Russell.

In the *Tractatus*, Wittgenstein (1922) clarified the notion of tautology by means of truth tables. Wittgenstein's own presentation of truth tables for the logical connectives appeared a bit clumsy, but they are just the table as shown in Fig. 5.1.

If we make the truth tables of the logical axioms in *Principia Mathematica*, we will observe that in all cases the final column in the table will contain nothing but T's. As an illustration we can consider the truth table for the axiom (2): $q \Rightarrow (p \vee q)$ (see Fig. 5.2).

That the truth value T appears in the whole final column means that independent of the truth values of p and q, the truth value of $q \Rightarrow (p \vee q)$ will be T. The axiom is true under all circumstances. This is what characterises a tautology, and this is what characterises the five axioms adopted in *Principia Mathematica*.

If the axioms are tautologies, then all theorems will be tautologies as well. In order to see this, one needs to look at the two rules of inference that were applied by Whitehead and Russell. One is the so-called Modus Ponens that has the following form:

If $A \Rightarrow B$ is a theorem, and A is a theorem, then B is also a theorem.

In order to understand that the quality of being a tautology is transferred from axioms to theorems due to the logical deduction, it is important to consider more carefully the Modus Ponens. If we assume that A is a tautology and that $A \Rightarrow B$ is a tautology, then B will also be a tautology.[10] Besides Modus Ponens, Whitehead and Russell apply a rule of substitution, roughly stating that one can substitute one symbol, such as p, with another symbol, such as q.

As all axioms in *Principia Mathematica* are tautologies and the property of being a tautology gets transferred from axioms to theorems by the two rules of deduction, Modus Ponens and substitution, then all the theorems in *Principia Mathematica* become tautologies as well. Furthermore, it can be proven that the system is complete as it contains as theorems all tautologies, and it is consistent as it contains as theorems nothing but tautologies. The axiomatisation of propositional logic as presented by Frege in the *Begriffsschrift* is complete and consistent as well.

By the clarification of the notion of tautology, we have reached a clarified expression of logicist absolutism: all mathematical theorems are tautologies. They are true under all circumstances.

Wittgenstein did send the *Tractatus* to Frege, who looked into it with reservations. He responded in a letter to Wittgenstein with some critical comments concerning the first page in the *Tractatus*. In a second letter Frege concentrated his critique on the first sentence in the *Tractatus*. There is no evidence that Frege ever looked further into the book. This is quite a sad story. Had Frege got as far as Wittgenstein's investigations of tautologies, he would have recognised a deep interpretation of his own basic claim that mathematical statements are logical statements.

The *Tractatus* was carefully studied by the Vienna Circle, and one of its members, Hans Hahn, commented on the claim that mathematics is nothing but tautologies. Hahn (1959) pointed out that it might be difficult to believe that the whole of mathematics is nothing but tautologies. However, he highlighted that there is one point that is overlooked, namely the fact that we human beings are not omniscient. An omniscient being would at once grasp all implications of any set of axioms, while we, human beings, have to move forward along a long row of deductions. The result might appear surprising to us; nevertheless, we are dealing with nothing but tautologies.

The situation, however, is more complex than outlined by Hahn. According to logicism, mathematics rests upon a foundation of logic, not only on the logic of propositional logic but also on that of predicate calculus, which operates with quantifiers. This implies that the notion of tautology needs to be further elaborated; it cannot just be interpreted with reference only to the truth tables. This has been carefully dealt with as part of the further elaborations of the logicist programme.[11]

[10] If $A \Rightarrow B$ is a tautology, then its truth-table will include only T's. If A is a tautology, then its truth-table will also include only T's. If B were not a tautology, then there will be a case where its truth value is F. This will mean that $A \Rightarrow B$ is not a tautology as assumed. Consequently, we must assume that B is a tautology as well.

[11] An important step was taken in 1930 by Gödel when he, as part of his Doctoral study supervised by Hahn, proved the completeness of the axioms of the functional calculus of logic. See Gödel (1983).

5.4 Extensional Logic

In his discussion of the notion of meaning, as presented in the article "Über Sinn und Bedeutung", first published in 1892, Frege (1969) made a distinction between *Sinn* and *Bedeutung*. He illustrated what he has in mind by referring to the notion of "morning star" and "evening star". The morning star is the star that appears most brilliant in the sky just before sunrise, while the evening star is the one that emerges just after sunset. However, in both cases we are referring to the exact same object, namely the planet Venus. The *Bedeutung* of the two notions "morning star" and "evening star" is the same, while their *Sinn* is different.

Normally the German word *Bedeutung* is translated as "reference", and the *Sinn* as "sense". According to Frege, the concept of meaning includes two components, namely reference and sense. The sense of "morning star" is different from the sense of "evening star", while the two expressions have the same reference. The two dimensions of meaning are similar to the distinction between the intention and extension of a concept.

In mathematics one finds many illustrations of this distinction, and Frege considered the notion "direction of a straight line *l*". The sense, the intention, of this notion can be clarified by considering where the line *l* might be pointing, while the reference, the extension, of the notion can be described as the set of lines parallel to 1. The sense, the intension, of a mathematical function *f* defined as $f(x) = x^2$, can clarified by explaining that the function *f* multiplies the input *x* by itself. The reference, the extension, of *f* is a set of order couples (x, y), where $y = x^2$.

Frege then considered what the sense and the reference of a statement could be, such as: there exist infinitely many prime numbers. One can try to clarify the sense of this statement by clarifying the notion of prime number and what could be meant by "infinitely many". The reference of "there exist infinitely many prime numbers" is, according to Frege, the truth value of the statement, which can be either *T* or *F*. The whole universe of possible references for any statement contains only two elements, namely *T* and *F*. This appears a most audacious claim, and so it is.

With this specification of "sense" and "reference", one can condense the whole outlook of logicism into the following: to understand what mathematics is, one must fully concentrate on the references of concepts and statements, and ignore the dimension of sense. Frege's harsh critique of psychologism, including all forms of empiricism, was that it tries to explore what mathematics is by clarifying the sense of some basic mathematical notions and statements. According to Frege, mathematics only concerns the reference of concepts and statements.

Frege's theory of meaning can be expressed in different ways in English, for instance as a distinction between connotation and denotation. The connotation of a concept *P* can be clarified in terms of the properties that it expressed by *P*. The denotation of the concept *P* is the set of objects *x* with these properties, i.e. the set $\{x|P(x)\}$, which we have also referred to as the extension of the concept *P*. The connotation of a proposition *p* is the fact that is stated by *p*. The denotation of the proposition *p* is the truth value of *p*. If what *p* is stating is in fact the case then the denotation of p is T; if what p is stating is in not the case then the denotation of p is F. Whatever statement we are dealing with, it has only two possible denotations, namely T or F.[12]

A logic focusing on the extensions of concepts and statements is referred to as an *extensional logic*, while a logic that considers the intensions of concepts is referred to as an *intensional logic*. (When making the distinction between extensional and intensional logic, it is common to spell the word "intensional" with the letter "s".) In an extensional logic the meaning of the logical connectives "and", "or", "if ... then", "if and only if", and "not" can be fully clarified by specifying how they operate on truth values. The logical connectives "and", "or", "if ... then", "if and only if" can be considered functions of two variables; for instance, one has $\vee(T, T) = T$; $\vee(T, F) = T$; $\vee(F, T) = T$; and $\vee(F, F) = F$. The logical connective "not" is a function of one variable, and we have $\neg(T) = F$, and $\neg(F) = T$. In extensional logic the functioning of the logical connectives can be fully clarified by truth tables.

An extensional language is language governed by extensional logic. In such a language, the truth value of a molecular statement is fully determined by the truth values of its atomic components. In order to determine the truth value of a molecular statement like.

$$(\mathrm{p} \vee (q \Rightarrow \neg r)) \Rightarrow ((r \vee s) \Rightarrow (p \vee \neg(s \vee r))$$

we only need to know the truth values of the atomic components p, q, r, and s. When these truth values are known, then the logical connectives determine the truth value of the molecular statement. A basic claim of logical positivism is that the language of science should be extensional. Consequently, mathematics can be nominated as the language of science.[13]

5.5 Limitations and Impacts

Logicism provided the philosophy of mathematics with a significant development. Never before has a philosophy of mathematics been elaborated in such technical detail.

During Modernity mathematics assumed new leading roles in human affairs. Mathematics developed as a central part of the scientific revolution, and since then more and more mathematical disciplines have opened up as further disciplines in the study of nature. Mathematics contributed to the Industrial Revolution and became

[12] I have also seen Frege's distinction between sense and reference as presented as being between meaning and reference. This I find less appropriate, as the distinction makes part of Frege's overall theory of meaning.

[13] See Sect. 2.5 for a discussion of the impact of this claim as expressed by logical positivism.

the quintessence of technological rationality. Mathematics played a number of different roles in the processes of globalisation, and now mathematics constitutes the digital technology behind global networking.

During Modernity, mathematics became broadly involved in scientific, technological, economic, and political affairs. Logicism developed as a philosophy of mathematics belonging to the Modern era. However, the logicist philosophy completely ignored how mathematics was involved in social affairs, as its all-determining ambition was to ensure that mathematical truths were, by necessity, truths. This would remove the socio-political critique of mathematics from the philosophic agenda. If things could not be different, critique became superfluous. Logicism contributed to the deification of mathematics by seeing it as a sublime expression of human rationality. It ignored the multitude of roles that mathematics was assuming in the modern project, and the fact that mathematics could play the role of hero as well as that of villain.

Logicism had an impact that reached far beyond the philosophy of mathematics. The whole discipline of formal logic developed from the logicist idea that logic has to serve as the foundation of mathematics. Logic has advanced into an independent scientific discipline, which has become crucial for computer science and digital technologies. The logicist ambition of turning logical deduction into patterns of calculation is brought into play in artificial intelligence.[14]

Logicism created the idea that the language of science should be extensional, meaning that it should concentrate on the references of concepts and statements. Mathematics was considered a paradigmatic example of an extensional language, and logicism directly inspired logical positivism to nominate mathematics as the language of science.

Logicism has had an impact on mathematics education. One of the central tenets of the Modern Mathematics Movement was the use of set theory, "right from the beginning!" The distinction between "sense" and "reference" is basic to Frege's theory of meaning. His audacious semantic claim was that mathematics should ignore the "sense" of concepts and statements and focus on their "references". Mathematics should fully concentrate on the extension of concepts—set theory—and on the extension of statements—their truth values. To a large extent, the Modern Mathematics Movement adopted Frege's theory of meaning by assuming that mathematics education should also ignore the senses of the concept to be learned, and fully concentrate on their references. This is the rationale behind "set theory right from the beginning".

The set-theoretical understanding of meaning brought about a particular approach for helping students understand complex, or molecular, mathematical concepts. In order for students to come to grasp a molecular concept, it was considered important that they first, step by step, came to understand the atomic parts of which the molecular concept was composed. For instance, in order to understand the concept

[14]In their discussion of artificial intelligence, McCulloch and Pitts (1990) refers explicitly to *Principia Mathematica*, Whitehead and Russell (1910/1913a/1913b).

of function, it was important to make the students understand the notion of set, of ordered pair, and of set of ordered pair. The teaching of a concept had to be organised according to the extensional decomposition of the concept. We are dealing with an explicit anti-holistic approach to the learning of mathematics.

The educational failure of the Modern Mathematics Movement has been broadly recognised. Again and again it has been pointed out that students did not find meaning in many of the activities in which they were supposed to participate. We are seeing here a grand-scale falsification of Frege's theory of meaning. The Modern Mathematics Movement assumed that an adequate language in mathematics education was extensional. But it is not. The set-theoretical theory of meaning is a semantic mistake with catastrophic consequences for mathematics education.

In spite of all its limitations, logicism is a philosophy of mathematics in ongoing development—although maintaining its narrow technical focus, and ignoring socio-political issues related to the development and application of mathematics.[15]

5.6 Cantor in Wonderland

While logicism operated at one end of the number line—being concerned with, for instance, the content of a statement like $1 + 2 = 3$—Georg Cantor (1845–1918) operated at the far-away other end of the line. He was captivated by the concept of infinity.[16]

Cantor paid particular attention to the notion of 1–1 correspondence. We call two sets A and B *equipotent* if and only if there exists 1–1 correspondence between them.

---

[15] Detailed investigations of Frege's philosophy of mathematic and language are found in Dummett (1981a, 1981b, 1991). Several have contributed to the further development of the logicist programme. Boolos (1998) engages in detailed studies of Frege's work, and he indicated how Frege can be saved from the paradox discovered by Russell. Burgess (2005) also bring Frege around Russell's paradox and outlined a neo-Fregean conception of mathematics. Hale and Wright (2005) bring together a range of studies establishing a neo-Fregean philosophy of mathematics, which were further synthetised by Hale and Wright (2005) in their study of logicism in the twenty-first century. Rayo (2005) outlines challenges for a neo-Fregean programme that he considered having got a life of its own independent of the original logicist programme. Rumfitt (1999) outlines a strategy for justifying the logicist approach for the arithmetic of the finite cardinal numbers. Wright (1983) sees numbers as objects, and in this way he expressed a Platonism similar to the one assumed by Frege. Zalta (1999) provides a reconstruction of features of Freges' *Grundgesetze der Arithmetik*, while Zalta, 2000) addresses metaphysical components included in the logicist programme.

[16] In 1895 Cantor published the original German version of "Contributions to the Founding of the Theory of Transfinite Numbers (First article)", and in 1897 the original German version of "Contributions to the Founding of the Theory of Transfinite Numbers (Second article)". These two articles brought together Cantor's principal contributions to the exploration of transfinite cardinal numbers. The two articles are found in Cantor (1955). Technical investigations of transfinite cardinal numbers are found in Cohen (2008). A philosophical investigation is presented in Tiles (2004).

A 1–1 correspondence is also referred to as a bijection, which is a function that is both injective and surjective.[17]

The set of natural numbers $N = \{1, 2, 3, \ldots.\}$ can be matched with the set of positive even numbers $E = \{2, 4, 6, \ldots\}$, meaning that a subset of natural numbers can be considered equipotent with the set of natural numbers itself.[18] The same observation can be made with respect to the set of prime numbers: the set of primes $P = \{2, 3, 5, 7, 11, \ldots\}$ is equipotent with the set N of natural numbers. The set of whole numbers Z is also equipotent with the set of natural numbers. What about the set of rational number Q? Does a bijection exist between N and Q? The answer is "yes", as the rational numbers can be listed in the following way:

$$Q = \{0, 1/1, -1/1, 1/2, -1/2, 2/1, -2/1, 1/3, -1/3, \ldots\}$$

There are repetitions in this list, but we can leave out the repetitions, and in this way recognise the 1–1 correspondence between N and Q.

The size of a set A is referred to as the cardinality of the set, and Cantor used the first letter in the Hebrew alphabet $\aleph_0$ with the index "0" to refer to the cardinality of the set of natural numbers. Thus we have:

$$Card(N) = Card(E) = Card(P) = Card(Z) = Card(Q) = \aleph_0$$

Such an observation could bring us to assume that when we get to infinity, everything becomes the same. The number line seems to end up in a big swamp of infinity. But Cantor showed that this is far from being the case. He opened up a route across this swamp and strode deep into a captivating Wonderland.

Let us start looking at a finite set A. The set of subsets of A is called the *powerset* of A, and it is symbolised by $\mathcal{P}(A)$:

$$\mathcal{P}(A) = \{x | x \subseteq A\}$$

If $A = \{1, 2, 3\}$, then $\mathcal{P}(A) = \{\emptyset, \{1\}, \{2\}, \{3\}, \{1,2\}, \{1, 3\}, \{2,3\}, \{1, 2, 3\}\}$. When $Card(A) = 3$, we have $Card(\mathcal{P}(A) = 2^3 = 8$. For any finite set A we have:

$$Card(\mathcal{P}(A)) = 2^{C(A)}$$

For any finite set A, the cardinality of the powerset $\mathcal{P}(A)$ is bigger than the cardinality of the set A. But does this also hold for infinite sets?

---

[17] This idea of equipotency was formulated by Hume in *A Treatise of Human Nature*, where he wrote: "When two numbers are so combined, as that the one has always a unit answering to every unit of the other, we pronounce them equal." (Hume, 2009, p. 121). In *Foundations of Arithmetic*, Frege referred explicitly to this statement by Hume (see, Frege, 1978, p. 73).

[18] Already in 1638 in *Two New Sciences*, Galilei (2000, pp. 40–41) expressed surprise at the fact that the set of natural numbers could be matched with the set of positive even numbers.

For showing that $Card(A) < Card(\mathcal{P}(A))$ also applies for an infinite set A, we need to specify what "<" might mean in this case. For two infinite sets A and B, $Card(A)$ is smaller than $Card(B)$, if a bijection exists between A and a subset of B, but does not exist between A and the whole set B. In other words, $Card(A) < Card(\mathcal{P}(A))$, if an injective function from A to $\mathcal{P}(A)$ exists, but no surjective function. Cantor then proved that for any set A, *finite or infinite,* we have

$$Card(A) < Card(\mathcal{P}(\mathrm{A}))$$

This proof created a whole Wonderland.[19]

Following Cantor, the cardinality of the set of natural numbers, $\aleph_0$, will be smaller than the Card($\mathcal{P}(N)$) which we refer to as $\aleph_1$. The cardinality of the powerset $\mathcal{P}(\mathcal{P}(N))$ of $\mathcal{P}(N)$ will be bigger than $\aleph_1$, so let us call it $\aleph_2$. And so we can continue. There exists an infinite sequence of bigger and bigger cardinal numbers:

$$\aleph_0, \aleph_1, \aleph_2, \aleph_3, \ldots$$

The numbers $\aleph_0$, $\aleph_1$, $\aleph_2$, $\aleph_3$, ... are also referred to as *transfinite cardinal numbers.*

Let us consider the set of real numbers R. Could one identify a 1–1 correspondence between N and R? Could $Card(N) = Card(R) = \aleph_0$? Cantor showed that this is not the case. In 1874 he published a proof, but in 1891, he published a different one in which he applied a "diagonal argument" that appeared to be a powerful way of arguing. To simplify Cantor's diagonal argument, we consider only the real numbers in the interval between 0 and 1, which we refer to as I, and we show that there does not exist a 1–1 correspondence between I and N. We start by assuming that such a 1–1 correspondence does exist, and show that this assumption leads to a contradiction. Consequently, we must conclude that the assumption was false, and that no such 1–1 correspondence does exist.

If the set of real numbers in the interval I is equipotent to the set of natural numbers, then set I could be enumerated in a list having the following form:

$$r_1, r_2, r_3, \ldots$$

[19]The proof runs as follows: We consider a set A, finite or infinite, and try to prove that there does *not* exist a surjective function f from A to $\mathcal{P}(A)$. We assume that f is a surjective function from A to $\mathcal{P}(A)$, and shows that this leads to a contradiction. We consider a set p belonging to $\mathcal{P}(A)$ defined in the following way: $p = \{x \in A | x \notin f(x)\}$. Since f is a surjective function, there is a x_0 of A such that $f(x_0) = p$. Does x_0 belong to the subset p of A? By definition of p, we have $x_0 \in p$ if and only if $x_0 \notin f(x_0)$. Since $f(x_0) = p$, we must conclude that $x_0 \notin p$. This means that $x_0 \in p$ if and only if $x_0 \notin p$. We have reached a contradiction and can conclude that there does not exist a surjective function from A to $\mathcal{P}(A)$.

In order to illustrate the argument, we assume that the start of the list is as follows:

$$r_1 = 0.\mathbf{5}0000000\ldots.$$
$$r_2 = 0.3\mathbf{3}333333\ldots.$$
$$r_3 = 0.17\mathbf{7}73333\ldots$$
$$r_4 = 0.234\mathbf{5}6789\ldots$$
$$r_5 = 0.7771\mathbf{1}177\ldots$$
$$r_6 = 0.00000\mathbf{0}11\ldots$$
$$r_7 = 0.343434\mathbf{3}4\ldots$$
$$\ldots$$

The point is to show that whenever we claim to have made a list of the real numbers in the interval I, there will be at least one real number from the interval I missing from the list. We call the missing number r_δ (δ for diagonal), and we show how it can be constructed. To identify the first digit in r_δ, we look at the first digit in r_1 which is 5. We then choose a digit different from 5; it could be 3, and we use 3 as the first digit in r_δ:

$$r_\delta = 0,3\ldots$$

For choosing the second digit in r_δ, we look at the second digit in r_2, which is 3. We then choose a digit different from 3; it could be 5, and we use 5 as the second digit in r_δ:

$$r_\delta = 0,35\ldots$$

We look at the third digit in r_3, which is 7, and we choose a digit different from 7; it could be 1, and use 1 as the third digit in r_δ, and we get

$$r_\delta = 0,351\ldots$$

We continue in this way: the fourth digit in r_δ should be different from the fourth digit in r_4, the fifth digit in r_δ should be different from the fifth digit in r_5, etc., and we get, say:

$$r_\delta = 0,3518256\ldots$$

According to this construction, r_δ which belongs to the interval I cannot be any of the numbers in the list: r_1, r_2, r_3, ... We have reached a contradiction: whichever list of real numbers from the interval I we are dealing with, it is possible to construct a real number r_δ also belonging to the interval I that is not in the list. In other words, there is no 1–1 correspondence between N and I, and consequently there is no 1–1 correspondence between N and R.

The *Card(R)* is bigger than $\aleph_0$, but could *Card(R)* = $\aleph_1$? The answer is "yes", as it can be proved that *Card(R)* = *Card*($\mathcal{P}(N)$).[20]

References

Bernays, P. (1983). On Platonism in mathematics. In J. van Heijenoort (Ed.), *From Frege to Gödel: A source book in mathematical logic, 1879–1931* (2nd ed., pp. 258–271). Cambridge University Press.

Boolos, G. (1998). *Logic, logic, and logic*. Harvard University Press.

Burgess, J. (2005). *Fixing Frege*. Princeton University Press.

Bussell, B. (1951). *The autobiography of Bertrand Russell*. Little, Brown and Company.

Cantor, G. (1955). *Contributions to the founding of the theory of transfinite numbers* (Translated, and provided with an introduction and notes by Philop E. B. Jourdain). Dover Publications.

Carnap, R. (1983). The logicist foundation of mathematics. In P. Benacerraf & H. Putnam (Eds.), *Philosophy of mathematics: Selected readings* (2nd ed., pp. 41–61). Cambridge University Press.

Cohen, P. J. (2008). *Set theory and the continuum hypothesis*. Dover Publications.

Dedekind, R. (1963). *Essays on the theory of numbers*. Dover Publications.

Doxiadis, A., Papadimitriou, C. K., Papadatos, A., & Donn, A. (2009). *Logicomix: An epic search for truth*. Bloomsbury.

Dummett, M. (1981a). *Frege: Philosophy of language*. Duckworth.

Dummett, M. (1981b). *The interpretation of Frege's philosophy*. Duckworth.

Dummett, M. (1991). *Frege: Philosophy of mathematics*. Duckworth.

Ernest, P. (2023). The ontological problems of mathematics and mathematics education. In M. A. V. Bicudo, B. Czarnocha, M. Rosa, & M. Marciniak (Eds.), *Ongoing advancements in philosophy of mathematics education* (pp. 3–41). Springer.

Frege, G. (1893/1903). *Grundgesetze der Arithmetik I-II*. H. Poble.

Frege, G. (1969). Über Sinn und Bedeutung. In G. Frege (Ed.), *Funktion, Begriff, Bedeutung: Fünf Logische Studien* (pp. 40–65). Vandenhoech & Ruprecht.

Frege, G. (1978). *Die Grundlagen der Arithmetik/The foundations of arithmetic*. Blackwell.

Frege, G. (1983). Begriffsschrift: A formula language, modelled upon that of arithmetic, for pure thought. In J. van Heijenoort (Ed.), *From Frege to Gödel: A source book in mathematical logic, 1879–1931* (2nd ed., pp. 1–82). Cambridge University Press.

Galilei, G. (2000). *Two new sciences* (S. Drake, Trans.). Wall and Emerson.

Gödel, K. (1983). The completeness of the axioms of the functional calculus of logic. In J. van Heijenoort (Ed.), *From Frege to Gödel: A source book in mathematical logic, 1879–1931* (2nd ed., pp. 582–591). Cambridge University Press.

Grassmann, H. (1861). *Lehrbuch der Arithmetikm für höhere Lehranstalten*. Verlag Adolph Enslin.

Hahn, H. (1959). Logic, mathematics and knowledge of nature. In A. J. Ayer (Ed.), *Logical positivism* (pp. 147–161). The Free Press.

Hale, B., & Wright, C. (2005). Logicism in the twenty-first century. In S. Shapiro (Ed.), *The Oxford handbook of philosophy of mathematics and logic* (pp. 166–202). Oxford University Press.

[20] A proof is based on the following idea: A real number r in the interval $I = [0;1]$, can be represented in a binary way as, say, $r = 0{,}110010011011100\ldots$ Any such sequence 110010011011100... can be associated with an element in $\mathcal{P}(N)$. From this we conclude: *Card* [0;1] = *Card* $\mathcal{P}(N)$). Based on this observation we can conclude *Card(R)* = *Card* $\mathcal{P}(N)$). See, for instance, Kleene (1971).

Heijenoort, J. van. (Ed.). (1967). *From Frege to Gödel: A source book in mathematical logic, 1879–1931*. Harvard University Press.

Hume, D. (2009). *A treatise of human nature*. The Floating Press.

Kitcher, P. (1983). *The nature of mathematical knowledge*. Oxford University Press.

Kleene, S. C. (1971). *Introduction to metamathematics*. Wolters-Noordhoff Publishing.

Maddy, P. (1997). *Naturalism in mathematics*. Oxford University Press.

Maddy, P. (2005). Three forms of naturalism. In S. Shapiro (Ed.), *The Oxford handbook of philosophy of mathematics and logic* (pp. 437–459). Oxford University Press.

McCulloch, W. S., & Pitts, W. H. (1990). A logical calculus of the ideas immanent in nervous systems. In M. A. Boden (Ed.), *The philosophy of artificial intelligence* (pp. 22–39). Oxford University Press.

Mill, J. S. (1970). *A system of logic*. Longman.

Peano, G. (1983). The principles of arithmetic, presented by a new method. In J. van Heijenoort (Ed.), *From Frege to Gödel: A source book in mathematical logic, 1879–1931* (2nd ed., pp. 83–97). Cambridge University Press.

Ravn, O., & Skovsmose, O. (2019). *Connecting humans to equations: A reinterpretation of the philosophy of mathematics*. Springer.

Rayo, A. (2005). Logicism reconsidered. In S. Shapiro (Ed.), *The Oxford handbook of philosophy of mathematics and logic* (pp. 203–235). Oxford University Press.

Rumfitt, I. (1999). Frege's logicism. *Proceedings of the Aristotelian Society, 73*, 151–180.

Russell, B. (1919). *Introduction to mathematical philosophy*. George Allen and Unwin.

Russell, B. (1951). *The autobiography of Bertrand Russell, 1872–1914*. Little, Brown and Company.

Russell, B. (1967). Mathematical logic as based on the theory of types. In J. van Heijenoort (Ed.), *From Frege to Gödel: A source book in mathematical logic, 1879–1931* (pp. 150–182). Cambridge University Press.

Tiles, M. (2004). *The philosophy of set theory: An historical introduction to Cantor's paradise*. Dover Publications.

Whitehead, A. & Russell, B. (1910/1913a/1913b). Principia mathematica I-III. Cambridge University Press.

Wittgenstein, L. (1922). *Tractatus logico-philophicus* (With an introduction by Bertrand Russell). Routledge and Kegan Paul.

Wright, C. (1983). *Frege's conception of number as objects*. Aberdeen University Press.

Zalta, E. N. (1999). Natural numbers as natural cardinals as abstract objects: A partial reconstruction of Frege's Grundgezetze in object theory. *Journal of Philosophical Logic, 28*(6), 616–660.

Zalta, E. N. (2000). Neologism? An ontological reduction of mathematics to metaphysics. *Erkenntnis, 53*(1–2), 19–65.

Chapter 6
Mathematics as Formal Structures

Abstract The formalist concept of mathematics was built in steps. The first step was taken by Hilbert when he investigated the foundation of geometry. It had long been recognised that Euclid's *Elements* had flaws, as some proofs were not only reached through logical deduction, but also based on intuitive readings of figures and diagrams. While Euclid presented five axioms, Hilbert presented 20 axioms as the foundations of geometry. A second step was taken by the metamathematical programme, which turned mathematical theories into objects for systematic study. They were concerned with the independence of mathematical axioms, the consistency and completeness of mathematical theories, and the possibility of solving the decision problem. A third step consisted of specifying what a formal system is in terms of: the alphabet the system; the sequences of symbols that count as formulas; the set of formulas that serve as axioms; the rules of inference to apply when making deductions; the notion of proof; and the definition of a theorem. With this clarification of a formal system, formalism declares that mathematics is made up of formalisms.

Keywords Formalism · Foundation of geometry · Metamathematics · Independence · Consistency · Completeness · Decision problem · Formal system · Alphabet · Formula · Axiom · Rule of inference · Proof · Theorem

David Hilbert has most often been referred to as the main contributor to the formalist philosophy of mathematics. However, this philosophy developed in steps, and very many people contributed to its development.[1]

Hilbert (1862–1943) was born in Königsberg, renamed as Kaliningrad in 1946. Through the city runs the river Pregolya which forms two islands. Königsberg had seven bridges: one connected the two islands, while the six others connected the two islands with the mainland. It was common to take a walk in the city, and it had turned into a public challenge to find a route where one passed all seven bridges, but only once. This turned out to be impossible in practice, and in 1735 Leonhard Euler

[1]The following presentation of formalism draws on Ravn and Skovsmose (2019) and on Skovsmose (in progress).

O. Skovsmose, *Critical Philosophy of Mathematics*, Advances in Mathematics Education, https://doi.org/10.1007/978-3-031-71375-0_6

(1707–1783) demonstrated that it was impossible in theory as well. Through this demonstration he founded topology as a mathematical discipline.

In 1880 Hilbert entered the University of Königsberg, and in 1885 he obtained a doctorate in mathematics. In 1895 he was appointed professor in mathematics at the University of Göttingen, and he remained there for the rest of his life.[2] Together with Felix Klein (1849–1925), who in 1886 had become a professor in Göttingen, Hilbert turned Göttingen into an important centre for mathematics research. In the latter part of his life, Klein became interested in mathematics education—or in "mathematical instruction" as it was called at the time. In 1908 he published the German version of *Elementary Mathematics from an Advanced Standpoint* (see Klein, 2016a, b, c), and in the same year he was elected president of the International Commission on Mathematical Instruction (ICMI).

The first step in developing the formalist philosophy of mathematics was taken when Hilbert investigated the foundation of geometry; a second step was completed by the formulation of the metamathematical programme; and the third consisted in the detailed and technical definition of a formal system. These steps led to the idea that mathematics is a formal structure.

6.1 The Foundation of Geometry

In the winter semester of 1897–1898, Hilbert decided to give a series of lectures about geometry. This came as a surprise to many of his students and colleagues, since previously he had not paid much attention to this topic. In 1899 the lectures were reworked and subsequently published in German. An English translation *Foundations of Geometry* (Hilbert, 2021) was published in 1902, which immediately became a bestseller in mathematical circles. Here, Hilbert concentrated on organising the foundations of geometry.

The classic axiomatisation of geometry is found in *Elements*, where Euclid laid the foundations using five axioms.[3] This axiomatisation soon became paradigmatic, not only to the presentation of mathematical knowledge, but to the presentation of any kind of knowledge. The axioms concerned relationships between well-defined entities, and Euclid started his exposition by defining the notions of point, line, and plane. The axioms of the system concerned the pre-defined notions, and they needed to be so simple that their absolute truth could be grasped immediately by intuition. It should be impossible to doubt any of the axioms. Using the axioms as a starting point and making deductions would lead to the proof of more complex geometric theorems, whose truth could not be grasped immediately, but whose *absolute* truth was guaranteed by the absolute truth of the axioms and by logical nature of the deduction.

[2]Biographic information about Hilbert can be found in Reid (1995).

[3]See Joyce (1998).

From the latter part of the nineteenth century onwards, intuition started to come under suspicion. It was considered one of the causes of paradoxes appearing in mathematics. Consequently, it became important to eliminate intuition from any form of mathematical reasoning. It seemed that certain proofs in *Elements* were not only reached through strict logical deduction, but also based on an intuitive reading of figures and diagrams. It came as a shock that there seemed to be flaws in Euclid's presentation—flaws that had not been noticed for more than 2000 years.

One of the intuitive insights that Euclid had taken for granted was that when a straight line intersects one of the sides of a triangle (and not in a corner of the triangle), then the straight line will also intersect one of the other sides of the triangle. A straight line cannot disappear within a triangle. No Bermuda triangle exists in geometry. Making a small sketch, one can easily feel convinced that the statement is true, and Euclid used it in his demonstrations. But the truth is based on an intuitive reading of a figure; it is not demonstrated to be a consequence of the five Euclidian axioms. This and other gaps in Euclid's axiomatisation were pointed out by Moritz Pasch (1843–1930), when he in 1882 published *Vorlesungen über neuere Geometrie*.[4]

Hilbert was aware of all such problems, and he wanted to eliminate any gaps in geometric proving. Hilbert's revision of the foundation of geometry included two components: one concerns the way that the basic notions are dealt with, another concerns the axioms that are applied.

Euclid started *Elements* with definitions of points, lines, and planes, but apparently he did not use these definitions in his demonstrations. Hilbert went further. He did not start *Foundation of Geometry* with any definition of the geometric notions, but just stated that he wanted to consider three different systems of "things". The things in the first system he called *points*; the things in the second system he called *lines*; and the things in the third system he called *planes*. Hilbert did not give any further clarification of the nature of points, lines, and planes, as he found that any question about the nature of points, lines, and planes was irrelevant for the development of geometry. What we know about the "things" in the three systems concerns their interrelationships as they are specified by the axioms. No other information is necessary for making geometric deductions. In this way Hilbert relegated ontological questions as being irrelevant for geometric theorising. Hilbert made this point clear by stating that instead of points, lines, and planes, one could

[4] See Pasch (1912).

also talk about tables, chairs, and beer mugs.[5] The elimination of intuition from geometry made geometric illustrations irrelevant.[6]

Hilbert was not the first to take an abstract view on geometry and abandon the relevance of defining points, lines, and planes. Hilbert had followed lectures by Herman Wiener (1857–1939), who presented an abstract view of geometry. A similar abstract view had been presented by Pasch and by Peano, but it was Hilbert that gave a clear formulation of the foundations of geometry without any ontological commitments.

While Euclid presented five axioms in *Elements*, Hilbert presented 20 axioms as providing the foundations of geometry. Hilbert organised the 20 axioms in five groups: (1) axioms of connection, (2) axioms of order, (3) axioms of parallels, (4) axioms of congruence, and (5) axioms of continuity.

Group (1) includes seven axioms, one being: "Two distinct points A and B always completely determines a straight line a."[7] Group (2) includes five axioms, one being: "If A, B, C are points of a straight line and B lies between A and C, then B lies also between C and A." This group also includes the axiom that a straight line cannot disappear within a triangle: "Let A, B, C be three points not lying in the same straight line and let a be a straight line lying in the plane ABC and not passing through any of the points A, B, C. Then, if the straight line a passes through a point of the segment AB, it will also pass through either a point of the segment BC or a point of the segment AC." This axiom is also referred to as Pasch's axiom.

Group (3)'s axioms of parallels include only one axiom: "In a plane α there can be drawn through a point A, lying outside of a straight line a, one and only one straight line which does not intersect the line a." The group (4) axioms of congruence include six axioms, one being: "If a segment AB is congruent to the segment $A'B'$ and also to the segment $A''B''$, then the segment $A'B'$ is congruent to the segment $A''B''$." Group 5 axioms include the so-called Archimedean axiom. In a simplified version, this axiom states that if we are dealing with two line segments, a smaller and a larger one, then if you add up the smaller one a sufficient number of times, the sum will be larger than the larger segments.

Besides the 20 geometric axioms, Hilbert added an axiom of completeness: "To a system of points, straight lines, and planes, it is impossible to add other elements in such a manner that the system thus generalized shall form a new geometry obeying

[5]It is not quite clear when Hilbert made this point. When quite young, before becoming profession in Göttingen, in a conversation with two other mathematicians at a train station he illustrated his point about the irrelevance of ontological issues with respect to geometry by referring to "tables", "chairs", and "beer mugs" (see Reid, 1995, p. 57). Shapiro (2000, p. 151) refers to a similar situation also taking place at a railway, although in 1891. Both references might well be true, as Hilbert's metaphorical use of "tables", "chairs", and "beer mugs" is extremely powerful, and he might have used it a number of times.

[6]In the *Foundations of Geometry,* Hilbert used many figures and diagrams. His way of doing has been carefully investigated in Hallett (2008).

[7]This and the following quotations are from Hilbert (2021).

all of the five groups of axioms". However, as he remarked, this axiom was "not of purely geometric nature".

After having presented the axiomatisation of geometry, Hilbert concentrated on showing the consistency of the system and the independence of the axioms. His strategy for showing the consistency of the axiomatisation was to apply analytic geometry to provide an arithmetic model of the axiomatisation. In this way he showed that the axiomatisation of geometry was consistent if the system of arithmetic was consistent. Keeping in mind that Felix Klein in 1871 had shown that non-Euclidian geometry was consistent if Euclidian geometry was consistent, one can see why it seemed urgent to Hilbert to secure the consistency of arithmetic. Hilbert's strategy for showing that all the axioms of the axiomatisation were independent—meaning that none of them could be proved from the others—was to use arithmetic models. He showed that a particular axiom was independent by presenting an arithmetic model where all the axioms were valid, except the one in question. In doing so it was demonstrated that this axiom could not be derived from the others.

6.2 Metamathematics

Hilbert was invited to deliver the opening plenary at the International Congress of Mathematicians to be held in 1900 at the University of Sorbonne in Paris. He was in doubt about what to say in his address, but one of his colleagues suggested that, instead of trying to summarise what had been done and achieved, he could try to look into the future and outline what could be done and achieved in the century to come. This turned out to be a splendid idea, and Hilbert gave an extremely important lecture that inspired mathematics research for many years to come.[8]

Hilbert pointed out 23 unsolved mathematical problems that were important to address. *Problem 1* he referred to as: "Cantor's problem of the cardinal number of the continuum". Hilbert was fascinated by the landscape of investigation that originated from Cantor's observation that the cardinality of the set of real numbers *R* is higher than the cardinality of the set of natural numbers *N*. *Problem 2* concerned: "The compatibility of the arithmetical axioms." Here Hilbert referred to a problem that emerged from his work with the *Foundation of Geometry*, where he presented a relative demonstration of the consistency of the axiomatisation of geometry. *Problem 10* concerned the determination of the solvability for the Diophantine equations. This problem led to the formulation of the more general decision problem.

Some of the problems instigated the conception of *metamathematics* as a study of mathematics. I will summarise the concern of metamathematics in terms of four clusters of problems. They concern the *independence* of mathematical axioms, the *consistency* of a mathematical theory, the *completeness* of a mathematical theory,

---

[8] An English translation of Hilbert's address can be found in Reid (1995, pp. 74–83).

and the *decision problem* concerning the possible existence of an algorithmic procedure for determining whether a given mathematical formula is a theorem or not in a given mathematical theory.

Independence concerns the question: Could one or some of the axioms be proved from the rest of the axioms? A classic question of independence concerns Euclid's fifth axiom. While the first four axioms in *Elements* appear simple, the fifth axiom has a higher degree of complexity. It states that, through a point outside a given line, it is possible to draw one and only one line parallel to the given line. (This is not the original formulation used by Euclid, but it is one of the many possible equivalents to this formulation.) Many attempts have been made to prove the fifth axiom from the remaining four. One approach was to negate the fifth axiom, and then make deductions from this negation, together with the four other axioms, until one reached a contradiction or an absurdity. Then one concluded that one could not negate the fifth axiom, which therefore was claimed to be proved from the other four axioms. There are two ways of negating the fifth axiom: one can assume that it is possible to draw more than one parallel line through a point outside the given line, or one can assume that it is not possible to draw any parallel lines through a point outside the given line. Both negations have been explored, and some strange implications have been observed. Some have claimed erroneously that the fifth axiom, therefore, had been proved. True enough, the implications appeared strange, but they were not illogical, and they could be considered theorems in another kind of geometry. It became recognised that it was possible to elaborate a deductive system, based on axioms that appeared false, but nevertheless were consistent. The possibility of formulating non-Euclidean geometries were first recognised by Friedrich Gauss (1777–1855), who kept this as a secret only to be shared with his closest colleagues. Nikolai Lobachevsky (1792–1856) and János Bolyai (1802–1860) elaborated non-Euclidean geometries in consistent ways.

Consistency concerns the question: Would it be possible from the chosen set of axioms to prove a false statement? Formulated differently: Would it be possible, through one deductive route, to show that a certain formula F is a theorem, and, through another deductive route, to show that $\neg F$ is a theorem? To demonstrate that an axiomatic system is consistent means to demonstrate that there exists no formula F, where one can derive both F and $\neg F$ as theorems in the system. The quest for consistency is crucial to metamathematics, as an inconsistent mathematical system is a disaster. It was crucial to Hilbert, by means of metamathematical studies to vaccinate mathematics for inconsistencies.

Completeness concerns the question: Could every true statement be proved as a theorem in the system? Formulated differently: Would it be possible for any formula F to demonstrate that either F or $\neg F$ is a theorem in the system? It had been assumed that Euclid's axiomatisation of geometry was complete, in the sense that all true geometric statements could be derived as theorems. Hilbert revealed that this was not the case; Euclid's axiomatisation of geometry was incomplete. To Hilbert the aim of metamathematics was to ensure that mathematical theories were both consistent and complete, and in this way to demonstrate the certainty of mathematical methods.

The *decision problem* concerns the question: Is it possible to identify, in an algorithmic way, whether a given formula F is a theorem in a given axiomatic theory or not? In other words: Is it possible to construct an algorithm that takes as its input a formula F from a mathematical theory, and as its possible output answers either "yes" or "no" to the question: Is F a theorem? The original German expression for the decision problem is *Entscheidungsproblem*, and this expression is often used in English texts as well. A specific case of the decision problem was presented by Hilbert as *Problem 10,* concerning the solution of Diophantine equations. In 1928, the decision problem was formulated in a more general way by Hilbert and Ackermann in the original German version of *Principles of Mathematical Logic* (Hilbert and Ackermann, 1950).[9]

In "The Foundation of Mathematics", based on a lecture given in 1927, Hilbert (1967) expressed his epistemic optimism is a programmatic way. His ambition was to eliminate "once and for all" uncertainties with respect to the foundation of mathematics. He wanted to do so by turning every mathematical proposition into a formula that can be "strictly derived". He wanted to recast mathematical definitions and deductions in a way where they would become "unshakable". In order to do so, he outlined a proof theory starting out with selected axioms, several of which were of pure logical nature. By means of this proof theory he wanted to show that "mathematics is a presuppositionless science, meaning that mathematics holds absolute truths."[10]

The issue of consistency was of particular importance to Hilbert. He was fascinated by Cantor's theory of transfinite cardinal numbers, which he also referred to as Cantor's Paradise. He wanted this paradise to be positioned on solid mathematical ground. In line with this ambition, Hilbert saw mathematics as composed of a finite part, including for instance traditional number theory, and an infinite part, including for instance the theory of transfinite cardinal numbers. The infinite mathematics included what Hilbert referred to as "idealised elements".[11] Hilbert wanted to demonstrate that no inconsistency would appear in the united system of finite and infinite mathematics. Consequently, the issue of consistency became urgent for Hilbert to solve. Furthermore, Hilbert wanted to prove the consistency in a way that could not be questioned, not even by those who did not find Cantor's theorising to be valid. Thus, Hilbert wanted to apply only "finite methods" when proving the

---

[9] For some mathematical theories the decision problem is solved, and as an example we can consider the formalised axiomatisation of propositional calculous (as presented in Sect. 5.3). It is demonstrated that a formula F is a theorem in this axiomatisation if and only if it is a tautology. Whether a formula F represents a tautology can be decided upon in an algorithmic way by elaborating its accompanying truth table. As a consequence, the decision problem can be considered solved for propositional calculous.

[10] For a careful study of Hilbert's program, see Kreisel (1983).

[11] See Hilbert (1983), which is an English translation of a lecture he gave in 1925. Here he discussed the notion of infinity and the relationship between what he referred to as "finitary statements" and "ideal statements".

consistency of the unified finite and infinite mathematics. This would bring about a demonstration that was compelling to everybody.

In his plenary in 1900 at the International Congress of Mathematicians, Hilbert expressed a radical version of epistemic optimism. It was his conviction that it was possible to solve any mathematical problem that was formulated in a precise way: "However unapproachable these problems may seem to us and however helpless we stand before them, we have, nevertheless, the firm conviction that their solution must follow by a finite number of purely logical processes."[12] In light of this optimism, it is not surprising that Hilbert did formulate the *Entscheidungsproblem*. That there might exist an algorithmic way of deciding whether or not a given formula is a theorem in a given formal theory, is a direct expression of the idea that it is possible to solve any mathematical problem "by a finite number of purely logical processes".

Metamathematics is a study of mathematics, but how are we to turn mathematics into an object for research? All kinds of mathematical theories exist, new theories are added, and many of them are in rapid and unpredictable development. How can we study such a wilderness? The metamathematical approach is to formalise mathematics, and to turn these formalisations into objects for investigation.

What methods can be applied in metamathematics? Metamathematics is the study of mathematics, so it appears circular to use mathematical methods in metamathematical studies. This is, in any case, what is taking place and also what Hilbert found necessary. However, he also pointed out important restrictions with respect to which methods could be allowed in metamathematics. Mathematics has applied a richness of patterns of reasoning, some of which might be more reliable than others. It is common to use indirect proof in mathematics, but since an aim of metamathematics was to demonstrate that indirect proof is valid in mathematics, such forms of reasoning could not be applied in metamathematics. An indirect proof of a formula F has the following form: one assumes $\neg F$, and shows that this leads to a contradiction; consequently, one claims that F has been proved. Hilbert suggested that metamathematics should only use *finite methods*, but he never managed to specify precisely what to consider a finite method.

In 1931, when Kurt Gödel published the original German version of *On formally undecidable propositions of Principia Mathematica and related systems*, the ambitions of metamathematics turned illusive. Gödel (1962) demonstrated that when a mathematical theory has a certain degree of complexity—so complex that it contains the theory of natural numbers—then it will be incomplete if it is consistent. This means that a consistent theory will always contain a formula F, where neither F nor $\neg F$ can be proved. Of course Hilbert wanted to use metamathematics to ensure both consistency and completeness of mathematical theories, but Gödel showed that this ambition was impossible to fulfil. In this way Gödel gave a serious blow to Hilbert's epistemic optimism. The knockout blow came the year after when Alan Turing (1937) showed that there is no general solution to the *Entscheidungsproblem*.[13]

[12] Quoted after Hilbert (1902, pp. 444–445). See also Reid (1995, p. 81).

[13] I will return to Turing's approach for solving the decision in Sect. 10.2.

Hilbert's epistemic optimism was a metaphysical conviction, but it was proven baseless by strict logical means.

At times, Gödel's incompleteness theorem is referred to as Gödel's *first* incompleteness theorem, as in *On formally undecidable propositions of Principia Mathematica and related systems* he dealt a second heavy blow to the metamathematical ambitions. He showed that, in order to demonstrate the consistency of a mathematical theory that is complex enough to contain the axiomatisation of the natural numbers, one needs to use metamathematical methods which are at least as complex as the methods used in the mathematical theory in question. Consequently, the ambition of restricting the metamathematical method to be finite is an illusion. This observation is at times referred to as Gödel's *second* incompleteness theorem.

6.3 Formalisms

In order to become an object for metamathematical investigation, mathematical theories need to be axiomatised and formalised. One could see formalisation as simply being preparation for metamathematical studies, but formalisations are more than that. Formalised mathematical theories can be seen as being *the real mathematics*. This idea defines the radical form of formalism as expressed by Haskell B. Curry (1951) in *Outlines of a Formalist Philosophy of Mathematics*.[14] It is not at all clear whether Hilbert himself assumed formalism in its radical form.

As an example of a formal system, we can consider a formalisation of propositional calculus such as that presented in *Principia Mathematica*. The first step in defining a formal system is to define the *alphabet* that is going to be used. The alphabet consists of the list of symbols that can be used; no other symbols are allowed except those from the pre-defined alphabet. The alphabet for propositional calculus could be:

$$(\,,\,),\neg,\ \vee\ ,\ \Rightarrow\ ,p,q,r,\ \ldots$$

One needs to imagine that infinitely many letters exist in the sequence $p, q, r, \ldots$ One could naturally also present the sequence as $p_1, p_2, p_3, \ldots$.[15]

The second step in defining a formal system is to state what counts as *formulas*. We need to set up rules for the "correct spelling" of strings of symbols. The rules could be of the following format:

[14] The ideas that Curry presented in *Outlines of a formalist philosophy of mathematics* was developed already 1939.

[15] More symbols can be introduced by definitions: $p \wedge q$ can be defined as $\neg(\neg p \vee \neg q)$; and $p \Leftrightarrow q$ can be defined as $(p \Rightarrow q) \wedge (q \Rightarrow p)$.

(1) All the variables p, q, r, ... are formulas.
(2) If A and B are formulas, then $(A \vee B)$ is a formula.
(3) If A and B are formulas, then $(A \Rightarrow B)$ is a formula.

It is taken as a given that no other strings of symbols count as formulas. The symbols A and B do not belong to the formal systems, but to the metalanguage by means of which the formal system is defined. This metalanguage also includes expressions like "all the variables", "are formulas", "if", and "then."

The third step is to present the *axioms* of the formal system. In *Principia Mathematica* the axioms for propositional calculus are:

(1) $(p \vee p) \Rightarrow p$
(2) $q \Rightarrow (p \vee q)$
(3) $(p \vee q) \Rightarrow (q \vee p)$
(4) $(p \vee (q \vee r)) \Rightarrow (q \vee (p \vee r))$
(5) $(q \Rightarrow r) \Rightarrow ((p \vee q) \Rightarrow (p \vee r))$

We can note that, for instance, the first axiom is written as $(p \vee p) \Rightarrow p$ and not as $((p \vee p) \Rightarrow p)$, as it should have been according to the definition of formulas. The convenient convention, however, is to omit the final set of parentheses.

The fourth step in defining a formal system it to list the *rules of inference* to be applied when making deductions. In *Principia Mathematica* two such rules are applied:

(1) If $A \Rightarrow B$ and A are theorems, then B is also a theorem.
(2) If A is a theorem, and B can appear from A by substituting one variable with another, then B is also a theorem.

Rule (1) is referred to as Modus Ponens, and the (2) as the Rule of Substitution.

The fifth step in defining a formal system is to specify what counts as a *proof*. A proof can be specified as a sequence of formulas $A_1, A_2, \ldots, A_n$, where any formula A_i $(1 \leq i \leq n)$ is either an axiom or can be derived from one or more of the formulas in the sequence $A_1, \ldots, A_{i-1}$ in accordance with the rules of inference.

The sixth step is to define what counts a *theorem* in the formal systems. A theorem can be defined as a formula that occurs as the last formula A_n in a proof of the form $A_1, \ldots, A_n$.

Taken together, the definitions of alphabet, formula, axiom, rule of inference, proof, and theorem specify what is meant by a formal system. Such a specification gives precision to the formalist definition of mathematics as being made up of formal systems.

In mathematical practice, and also in practices that observe formalism, not all of the six steps are completely dealt with. For instance, the rules of inference are normally not explicitly stated in mathematical theory building. The proofs are "just" observing general logical rigour. Formalism exists in different versions, and not all of them embrace the definition of formal structure that I have just outlined.

As an illustration of this, I will make some remarks about group theory. A group $(G, \bullet)$ is composed of a set G containing the elements $a, b, c, \ldots$, and a function $\bullet$ of two variables from the set $G \times G$ to the set G. Instead of writing $\bullet(a, b)$, we write $a \bullet b$. The axioms are:

(1) $\forall a, b, c \in G : (a \bullet b) \bullet c = a \bullet (b \bullet a)$
(2) $\exists n \in G : \ \forall a \in G : a \bullet n = n \bullet a = a$
(3) $\forall a \in G : \ \exists a^{-1} \in G : a \bullet a^{-1} = a^{-1} \bullet a = n.$

The axiom (2) refers to a neutral element n from G, having the property $a \bullet n = n \bullet a = a$ for all elements a from G. The symbol a^{-1} used in axiom (3) refers to the inverse elements of a defined by $a^{-1} \bullet \mathrm{a} = \mathrm{a} \bullet a^{-1} = n$. One of the first theorems to be proved in group theory is that there exists only one neutral element n. The proof starts by assuming that two neutral elements n_1 and n_2 exist. According to the definition we have $n_1 = a \bullet a^{-1}$, and we also have $n_2 = a \bullet a^{-1}$. This brings us to the equation $n_1 = a \bullet a^{-1} = n_2$, and we see that $n_1 = n_1$. We have made the initial proof in group theory according to a normal mathematical practice. However, we have not outlined any rules of inference. This is often done when we are dealing with formal logic, but it is rarely seen when we are dealing with other mathematical theories, like group theory. Still, we can see the presentation of group theory as an example of a formalist presentation of mathematics.

The formalist understanding of mathematics answers some of the classic philosophic questions such as: What is the nature of mathematical entities? What do notions like point, line, and plane refer to? What do notions like natural number, negative number, real number, and complex number refer to? According to formalism, any such ontological questions become irrelevant. The basic concepts of a formalised theory do not need any specification. In the definition of a group $(G, \bullet)$, G is a set containing the elements $a, b, c, \ldots$ It is irrelevant to ask about the nature of these elements. The only important thing is how they relate to each other. This observation does not only apply to group theory, but to any formalised theory. Mathematics is just about "things", and we can label these "things" whatever way is convenient.

Formalism also seems to provide an answer to the question: What is the nature of mathematical truths? Are mathematical truths necessary truths, or are they basically empirical truths? According to formalism one needs to distinguish between two kinds of truths: one concerns the truth of a mathematical statement, the other concerns the truth of a mathematical theory. The truth of a mathematical statement concerns its provability within an axiomatic system. A mathematical statement is true, relative to a formal system, if and only if it can be derived as a theorem within the system. Truths of mathematical statements are not absolute truths; they are relative truths. They have an if-then format: if the axioms are true, then the deduced theorems are true as well. Formalism advocates an *if-then absolutism* with respect to mathematical statements. The situation is quite different with respect to the truth of a mathematical theory. A mathematical theory can be applied in order to describe some empirical phenomena. It can provide more or less precise and reliable descriptions. The degree to which it does so is an empirical question. Formalism advocates an empirical graduation of truth with respect to mathematical theories.[16]

6.4 Limitations and Impacts

Formalism provides a most elaborate technical interpretation of mathematics, but it is also a limited interpretation. Formalism focuses sharply on mathematical concepts and theories themselves, but it is oblivious to the social context of mathematical theory building and to the possible impact of bringing mathematics into action. Formalism treats mathematics as detached from any kind of social reality.

Different from logicism, formalism has turned into a broadly assumed working philosophy of mathematics. Here I refer to a conception of mathematics that need not be expressed in an explicit way, but provides indirect guiding of mathematical research practice. As a working philosophy, formalism can create the image that mathematics is a kind of game, where one explores the logical impact of selected axioms. This makes mathematical research appear to be an innocent and harmless activity, principally guided by unsolved intrinsic mathematical problem. That mathematics research might be guided by external demands—being of an economic, political or military nature—becomes camouflaged by the game metaphor. Formalism establishes an easy-going discourse that turns mathematics into a sublime and "pure" logical activity.

As a working philosophy of mathematics, formalism inserts mathematics into an ethical vacuum. By its very nature, mathematics is claimed to be neutral: it does not reflect any political priorities, and it is not involved in any controversial issues. I use ethics in a broad sense here as referring to any kind of socio-political reflections of bringing mathematics into action. However, in being presented as a logical game, such ethical reflections become irrelevant. The formalist understanding of mathematics has been repeated again and again, and has become part of a broadly applied discourse about mathematics. Mathematics is described as a detached activity: it is not formed by external social forces, nor does it have an impact on social affairs.

Despite its internal focus, formalism has had a tremendous impact on technological developments. Hilbert formulated the proof theory in an attempt to show that any well-defined mathematical problem has a solution. The proof theory further developed the formalisation of logic as initiated by Frege and further elaborated by Whitehead and Russell. The epistemic optimism that Hilbert associated to the metamathematical program was ruined by the insight provided by Gödel and Turing, but the techniques developed as part of proof theory made artificial intelligence

---

[16] Many contributes to the formalist programme today. Azzouni (2005) discusses ways for formalism to escape Platonism. Detlefsen (2005) gives an account to the historical roots of the formalist ideas. Gabbay (2010) specifies a formalist philosophy of mathematics with particular reference to arithmetic. Weir (2016) investigates relationships between informal and formal proving.

feasible. The creation of artificial intelligence is grounded in formalisations of logical procedures.[17]

The Bourbaki group, which was established in the mid-1930s, assumed a formalist view of mathematics which they elaborated further into a structuralist view. At the beginning the members of the Bourbaki group were mainly French mathematicians, but over time many others joined the group. The original aim was to provide an updated presentation of mathematical analyses, making use of new and more general terminology, thereby adding to the conceptual rigour of mathematics. However, soon the aim changed, and became to provide a systematic presentation of major parts of mathematics. The results appeared in *Éléments de Mathématique,* the first volume of which was published in 1939 (see Bourbaki, 2004a, b, c).[18]

According to Bourbaki it is possible to identify some basic mathematical structures, such as the group $(G, \bullet)$.[19] The group structure can be identified in a variety of mathematical disciplines: number theory, geometry, and vector calculus. By means of these basic mathematical structures, Bourbaki showed how other mathematical structures can be defined, such as ring, field, ordered field, vector space, and Hilbert space. The Bourbaki architecture of mathematics is a composition of overlapping structures, and set theory provides the base for defining such structures. The Bourbaki group was fully concentrated on presenting the structural architecture of mathematics and did not pay any attention to possible applications of mathematics. This was considered irrelevant for identifying mathematical structures. The historical development of mathematical ideas, concepts, and theories was also considered irrelevant for creating the structural view of mathematics. The Bourbaki architecture gives a synchronic presentation of mathematics.

Bourbaki's structuralism became a principal source of inspiration for the Modern Mathematics Movement that, during the 1960s and 1970s, had a strong impact on mathematics education the world around. This structuralism also inserted mathematics education in the ethical vacuum, together with other mathematical practices.

6.5 Enjoying the Wonderland

Georg Cantor showed that the cardinality of the set of real numbers is bigger than the cardinality of the set of natural numbers. He used the symbolism Card $(N) = \aleph_0$ and Card$(R) = \aleph_1$, so his observation was:

$$\aleph_0 < \aleph_1$$

[17] For an overview of the philosophy of artificial intelligence, see Boden (2005).

[18] For a systematics presentation of structuralism in mathematics, see Hellmann (2005).

[19] Bourbaki operates with three mother structures: a set organised by an operation; a set organised by a relation; and a set organised by a topology.

As soon as we have recognised that $\aleph_0$ and $\aleph_1$ are different, we can ask if there might exist an infinity somewhere in between them. Cantor did not think so, and this belief is referred to as the *continuum hypothesis*. However, Cantor could not prove the hypothesis.

In order to see how the continuum hypothesis was addressed, it is necessary to say something about set theory. The paradox observed by Russell showed the existence of a flaw in Frege's conception of the formation of sets. Frege was too liberal, meaning that too many sets could be generated, which made it possible for set-theoretical paradoxes to occur. Whitehead and Russell solved the problem by introducing a theory of types, which restricted the generation of sets. The implication, however, was that too few sets were generated for fulfilling what was needed in mathematics.

In 1908, Ernst Zermelo proposed a solution by means of an axiomatisation of set theory (see Zermelo, 1967). Zermelo had become inspired to work with set theoretical issues by Hilbert's presentation in 1900 of the 23 problems, where the first problem was formulated as: "Cantor's problem of the cardinal number of the continuum". Zermelo was not able to demonstrate that his axiomatisation of set theory was consistent, but this was accomplished by Abraham Fraenkel, and it is common to talk about the Zermelo-Fraenkel axiomatisation of set theory, also referred to as the ZF axiomatisation. The axiom of choice has been added to the system, and then we talk about ZFC axiomatisation.

There are nine axioms in ZFC, and I will mention three of them, just to give an impression of the system. The axiom of extensionality states that two sets are equal if they have the same elements. The axiom of powerset states that for any set A, it is possible to form a set of all the subsets of A. And the axiom of choice states that if we have an infinite sequence of sets $A_1, A_2, A_3, \ldots$, then it is possible to form a new set by picking an element from each for the sets $A_1, A_2, A_3, \ldots$

Kurt Gödel (1940) showed that Cantor's continuum hypothesis cannot be disproved within the ZFC axiomatic system, and Paul J. Cohen (1963) showed that the continuum hypothesis cannot be proved in the ZFC axiomatic system. We might ask whether or not there exists a set with a cardinality bigger than $\aleph_0$, but smaller than $\aleph_1$. However, it seems this is not a yes-no question. We are free to choose the answer. If we want, we can add, as an additional axiom to ZFC, that there exists a set with such a cardinality between $\aleph_0$ and $\aleph_1$. We can also add as an additional axiom that such a set does not exist. With reference to ZFC, we are free to assume whether the continuum hypothesis is valid or not. This gives an answer to the continuum hypothesis, although not an answer that was expected.

But the answer seems strange. Should we not assume that *in reality* a set with a cardinality between $\aleph_0$, and $\aleph_1$ either exists or does not exist? Should we not assume that it is some limitation of the axiomatisation of set theory that makes space for us to choose? Should we not assume that formalisation of set theory does not fully depict the set-theoretical reality? But what reality are we talking about?

Let us assume that we have added an axiom to ZFC stating that a set with a cardinality $\aleph_x$ exists where:

$$\aleph_0 < \aleph_x < \aleph_1$$

Are we also free to assume that a set with cardinality $\aleph_y$ exists, where:

$$\aleph_0 < \aleph_y < \aleph_x$$

And could we continue? Are we free to choose whether there are infinitely many different cardinal numbers between $\aleph_0$ and $\aleph_1$?

Maybe by now we have completely lost contact with any kind of mathematical reality. But we are enjoying the Wonderland.

References

Azzouni, J. (2005). How to nominalize formalism. *Philosophia Mathematica, 13*(2), 135–159.

Boden, M. A. (Ed.). (2005). *The philosophy of artificial intelligence.* Oxford University Press.

Bourbaki, N. (2004a). *Elements of mathematics: Elementary theory.* Springer.

Bourbaki, N. (2004b). *Elements of mathematics: Set theory.* Springer.

Bourbaki, N. (2004c). Elements of mathematics: Functions of real variable. In *Springer. Selected readings* (2nd ed., pp. 202–206). Cambridge University Press.

Cohen, P. J. (1963). The independence of the continuum hypothesis. *Proceedings of the National Academy of Sciences of the United States of America, 50*(6), 1143–1148.

Curry, H. B. (1951). *Outlines of a formalist philosophy of mathematics.* North-Holland Publishing Company.

Detlefsen, M. (2005). Formalism. In S. Shapiro (Ed.), *The Oxford handbook of philosophy of mathematics and logic* (pp. 236–317). Oxford University Press.

Gabbay, M. (2010). A formalist philosophy of mathematics, Part I: Arithmetic. *Studia Logica, 96*, 219–238.

Gödel, K. (1940). *The consistency of the axiom of choice and of the generalized continuum hypothesis with the Axioms of Set Theory.* Princeton University Press.

Gödel, K. C. (1962). *On formally undecidable propositions of principia mathematica and related systems.* Basic Book.

Hallett, M. (2008). Reflections on the purity of methods if Hilbert's *Grundlagen der Geometrie.* In P. Mancosu (Ed.), *The philosophy of mathematical practice* (pp. 198–255). Oxford University Press.

Hellmann, G. (2005). Structuralism. In S. Shapiro (Ed.), *The Oxford handbook of philosophy of mathematics and logic* (pp. 536–562). Oxford University Press.

Hilbert, D. (1902). Mathematical problems. *Bulletin of the American Mathematical Society, 8*, 437–479.

Hilbert, D. (1967). The foundations of mathematics. In J. van Heijenoort (Ed.), *From Frege to Gödel: A source book in mathematical logic, 1879–1931* (pp. 464–479). Harvard University Press.

Hilbert, D. (1983). On the infinity. In P. Benacerraf & H. Putnam (Eds.), *Philosophy of mathematics: Selected readings* (2nd ed., pp. 183–201). Cambridge University Press.

Hilbert, D. (2021). *Foundations of geometry* (E. J. Townsend, Trans.). Altus Classics.

Hilbert, D., & Ackermann, W. (1950). *Principles of mathematical logic.* Chelsea.

Joyce, D. E. (1998). *Euclid's elements.* http://aleph0.clarku.edu/~djoyce/java/elements/elements.html
Klein, F. (2016a). *Elementary mathematics from an advanced standpoint. Volume I: Arithmetic, algebra, analysis.* Springer.
Klein, F. (2016b). *Elementary mathematics from an advanced standpoint. Volume II: Geometry.* Springer.
Klein, F. (2016c). *Elementary mathematics from an advanced standpoint. Volume III: Precision mathematics and approximation mathematics.* Springer.
Kreisel, G. (1983). Hilbert's programme. In P. Benacerraf & H. Putnam (Eds.), *Philosophy of mathematics: Selected readings* (pp. 207–238). Second Edition. Cambridge University Press.
Pasch, M. (1912). *Vorlesungen über neuere Geometrie.* B.G. Teubner.
Ravn, O., & Skovsmose, O. (2019). *Connecting humans to equations: A reinterpretation of the philosophy of mathematics.* Springer.
Reid, C. (1995). *Hilbert.* Springer.
Shapiro, S. (2000). *Thinking about mathematics: The philosophy of mathematics.* Oxford University Press.
Skovsmose, O. (in progress). Formalism, structuralism, and the doctrine of neutrality. In B. Greer, D. Kollosche, & O. Skovsmose (Eds.), *Breaking images: Iconoclastic analyses of mathematics and its education.* Open Book Publishers.
Turing, A. M. (1937). On computable numbers, with an application to the Entscheidungsproblem. *Proceedings of the London Mathematical Society, s2-42*(1), 230–265.
Weir, A. (2016). Informal proof, formal proof and formalism. *The Review of Symbolic Logic, 9*, 23–43.
Zermelo, E. (1967). Investigations in the foundations of set theory I. In J. van Heijenoort (Ed.), *From Frege to Gödel: A source book in mathematical logic, 1879–1931* (pp. 199–215). Harvard University Press.

Chapter 7
Mathematics as Mental Acts

Abstract Intuitionism accuses formalism of confusing mathematics with formal systems. Mathematics is a metal act, and not a string of symbols. According to intuitionism, many of the proofs used in mathematics are not valid. The paradoxes that brought mathematics into a foundational crisis would all evaporate if the process of proof had been "constructive". This brings intuitionism to articulate a logic that is different from classical logic; for instance, the principle of the excluded middle is not valid in intuitionistic logic. In classical logic the meaning of the logical connectives can be clarified in terms of truth tables, but this is not so in intuitionistic logic. Through a discussion of language, mathematics, and meaning, Brouwer inspired Wittgenstein to return to philosophy, and he inspired Freudenthal to see mathematics as a human activity.

Keywords Three magi · Topology · Brouwer's fixed-point theorem · Intuitionism · Mathematics as mental act · *Formen der Anschauung* · Formalism · Logicism · Constructive proof · Principle of the excluded middle · Classic logic · Logical connective · Truth table · Mathematics as a human activity · Modern mathematics movement · Wonderland

Intuitionism is the third of Three Magi, the philosophies worshipping mathematics. Like logicism and formalism, it also suffers from a profound blind-spot in relation to the way that mathematics is involved in social affairs.

The intuitionist philosophy of mathematics was formulated by Luitzen Egbertus Jan Brouwer (1881–1966). When he was 16, he enrolled at the University of Amsterdam and registered as a student in the Faculty of Mathematics and Science. Brouwer had a broad set of interests, and he was not particularly focused on mathematics. The young Brouwer's reflections on life and death were expressed in *Life, Art and Mysticism*, first published in 1905. In this booklet, full of metaphysical speculations, Brouwer (1996) also expressed an epistemic absolutism:

> Always and everywhere truth is in the air, and whenever it breaks through, truth will always be the same for those who understand. (p. 404)

O. Skovsmose, *Critical Philosophy of Mathematics*, Advances in Mathematics Education, https://doi.org/10.1007/978-3-031-71375-0_7

Many of the reflections in *Life, Art and Mysticism* stayed in the booklet, but Brouwer brought epistemic absolutism with him into intuitionism.

In his biography *L. E. J. Brouwer—Topologist, Intuitionist, Philosopher: How Mathematics Is Routed in Life*, Dirk van Dalen (2013) states that for several years Brouwer's relationship with mathematics remained ambiguous and that it was only when he had success in topology that he "definitely resign himself to a mathematician's life" (p. 44). Brouwer contributed to the further development of topology, which had been anticipated as a mathematical discipline by Leonhard Euler in 1735, when he solved the Königsberg problem. In topology Brouwer proved a fixed-point theorem, which can be illustrated in the following way: in a cup of coffee, all the coffee particles are in a certain position. When stirred by a teaspoon, the particles will move into new positions. But not all of them! According to the fixed-point theorem there will be at least one particle that will occupy the same position as it did before the stirring.[1]

When Brouwer had "resigned" himself to mathematics, he started developing intuitionism both as a mathematical approach and as a philosophy of mathematics. While logicism claimed mathematics to be logic and formalism claimed it to be formal structures, intuitionism found mathematics to be made up of mental acts.[2]

7.1 Intuitionism versus Formalism

In 1909 Brouwer gained a post the University of Amsterdam, and in 1912 he was appointed professor. He gave an inaugural lecture with the title "Intuitionism and Formalism", where he presented the intuitionist programme.[3]

Brouwer (1983) profiled intuitionism by confronting it with formalism, which at the time was being developed considerably, guided by David Hilbert's ideas. Formalism could be associated with the Mathematical Department at Göttingen University, which together with the Mathematical Department at the University of Berlin could be considered the world's most important centre for mathematics research. Brouwer established the Mathematical Department at the University of Amsterdam as the centre for intuitionistic mathematics. In "Intuitionism and Formalism", Brouwer did not show any particular interest in logicism, nor did he do so in his later writings. He found that the principal divergence in the conception of mathematics was between intuitionism and formalism.

[1]In mathematical terms the fixed-point theorem states that for any continuous function f mapping a compact convex set onto itself, there exists a point x_0 such that $f(x_0) = x_0$. One can naturally question if the stirring with a spoon represents a continuous function, but the cup of coffee only serves as an illustration.

[2]In the following, I draw in particular on Dalen (2013), Ravn and Skovsmose (2019), and Skovsmose (in progress). See also Pambuccian (2000).

[3]The lecture waspublished in 1913 in the *Bulleting of the American Mathematical Society* and reprinted in Benacerraf and Putnam (1983, pp.77–89).

Brouwer started the lecture by pointing out that he wanted to address the foundation of mathematics, which at the time was a crucial issue for the philosophy of mathematics. A broadly assumed view was that the remarkable usefulness of mathematics for describing natural phenomena was rooted in the fact that nature operates according to mathematical patterns, and that by means of mathematics one can capture the laws of nature, as they as such have a mathematical format. This is, however, not the way Brouwer saw the situation. According to him, mathematics is not a reflection of the order of nature; instead, mathematics embodies the order of the human mind. By means of mathematics, human beings create order in what they perceive, including natural phenomena.

Brouwer (1983) stated that "man always and everywhere creates order in nature" (p. 77). In making this formulation, Brouwer embraced a basic idea from Immanuel Kant's *Critique of Pure Reason*. In this work, Kant (1973) tried to grasp the nature of knowledge including mathematical knowledge: Why can one associate certainty with mathematical propositions? How can it be that mathematics can be applied to natural phenomena and express natural laws which demonstrate certainty? In answering these questions, Kant made what has been referred to as a Copernican revolution with respect to the understanding of the relationship between mathematics and reality. Kant confronted any version of empiricism according to which all knowledge emerges from sense impressions. This epistemic idea is clearly formulated by, for instance, John Locke (1997), who saw the human mind as originally being a *tabula rasa* that over time becomes filled with more and more inscriptions in terms of sense impressions. In the end these inscriptions add up to becoming human knowledge. While Locke presented the human mind as a passive *tabula rasa*, Kant gave the human mind an active role in the formation of knowledge. According to Kant, we, human beings, are always operating with forms of perception, *Formen der Anschauung*, which are "epistemic categories". We cannot dispense with such categories; they operate *a priori* to the formation of any kind of knowledge. We do not passively receive sense impressions that add to our knowledge about the world, but our forms of perception format what we perceive and how we perceive it. What we perceive is not copies of a pre-existing reality; instead, *Formen der Anschauung* organises our perceptions.

According to Kant, time and space are two basic *Formen der Anschauung*. Whatever phenomenon we experience takes place in time and space. This is not a particular empirical insight; it is a precondition for any kind of insight. Thus time and space do not refer to some empirical phenomena, but to structures of the human mind. They are containers by means of which we grasp and understand what we perceive. Following Kant, the basic structures of time and space are expressed by mathematics: one can see arithmetic as capturing the structure of time, while geometry captures the structure of space. It is us, human beings, who project mathematics onto our experiences. As a consequence, mathematics necessarily applies to our experiences.

The German word *Anschauung* is often translated as *intuition*, and it is precisely due to the active role of intuition, *Formen der Anschauung*, in the formation of knowledge that Brouwer (1983) referred to Kant as representing intuitionism.

However, Brouwer found that Kant represented an old form of intuitionism, which it is important to leave behind in order to develop what he referred to as neo-intuitionism. I will not, however, use the expression neo-intuitionism when referring to Brouwer, but instead just intuitionism, as has become common.

One might associate intuition with some degree of uncertainty and vagueness. This was how intuition was seen by formalism, which found that the use of intuition created space for paradoxes to emerge. Consequently, formalism found that intuition had to be eliminated from mathematics. This was not Brouwer's opinion. Like Kant, Brouwer did not associate doubt and vagueness with *Anschauung* or intuition. To him intuition ensures exactness in mathematical thinking and constitutes mathematical truths as necessary truths.

Brouwer was inspired by Kant, but was also worried about Kant's old form of intuitionism, which left the intuitionist opposition to formalism weak: "As long as the intuitionists adhered to the theory of Kant it seemed that the development of mathematics in the nineteenth century put them in an ever-weaker position with regard to the formalists" (p. 79). A big problem for the old form of intuitionism was the discovery of non-Euclidian geometry, which could be interpreted as a direct falsification of Kant's claim that space is organised *a priori* according to the Euclidian geometry. Brouwer found that the discovery of non-Euclidean geometry was the most serious blow to the Kantian theory, but not to intuitionism in its new format.

While Kant operated with time and space as two parallel *Formen der Anschauung*, Brouwer concentrated on the intuition of time:

> [...] neo-intuitionism considers the falling apart of moments of life into qualitatively different parts, to be reunited only while remaining separated by time as the fundamental phenomenon of the human intellect, passing by abstracting from its emotional content into the fundamental phenomenon of mathematical thinking, the intuition of the bare two-oneness. This intuition of two-oneness, the basal intuition of mathematics, creates not only the numbers one and two, but also all finite ordinal numbers, inasmuch as one of the elements of the two-oneness may be thought of as a new two-oneness, which process may be, repeated indefinitely; this gives rise still further to the smallest infinite ordinal number ω. (p. 80).

Brouwer saw time as a "falling apart of moments", as the principal phenomenon in which mathematical thinking is rooted. To Brouwer, time was not an empirical notion, but an abstract epistemic category stripped of any emotional content.

A crucial notion in this condensed formulation of the intuitionist conception of mathematics is the *two-oneness*. What could Brouwer mean by this expression? In the original Dutch version, Brouwer used the expression *twee-enigheid.* It is an expression with religious connotations, as it can be associated with the notion of the Trinity. Thus, the Dutch word for trinity is *drie-eenheid.* If one kept closer to Brouwer's Dutch terminology, then the expression *twee-enigheid* would be translated as *twoity*. What Brouwer wanted to highlight by the expression *two-oneness* is that counting is the basic mathematical act.

Counting is a discrete mathematical process, while much mathematics, geometry for instance, operates with continuous phenomena. How could all kinds of mathematics be rooted in the discrete process of counting? This is a huge challenge for intuitionism, but in "Intuitionism and Formalism", Brouwer pointed out that an intuition of space is not necessary for the intuitionist programme, but that the intuition of time also grasps phenomena related to geometry. Brouwer pointed out that "the apriority of time does not only qualify the properties of arithmetic as synthetic a priori judgments, but it does the same for those of geometry, and not only for elementary two- and three-dimensional geometry, but for non-Euclidean and n-dimensional geometries as well" (p. 80).

With the problem of non-Euclidian geometry pushed aside, and with the insight that the basic intuitive act is counting, Brouwer was ready to confront formalism. According to Brouwer (1983), formalism has no conception of what a mathematical activity does include, which gives rise to the erroneous conviction that mathematical exactness is established through a formal game with symbols:

> For the formalist [...] mathematical exactness consists merely in the method of developing the series of relations, and is independent of the significance one might want to give to the relations or the entities which they relate. And for the consistent formalist these meaningless series of relations to which mathematics are reduced have mathematical existence only when they have been represented in spoken or written language together with the mathematical-logical laws upon which their development depends, thus forming what is called symbolic logic. (p. 78–79).

Brouwer could only distance himself from the formalist dogma by arguing that mathematical certainty and exactness have an if-then format, rooted in a game with logical symbols. To Brouwer the formalist misunderstanding of mathematics was expressed by Hilbert stating that the nature of mathematical entities is irrelevant to mathematics. The formalist misunderstanding is compressed in Hilbert's claim that instead of points, lines, and planes one might as well talk about tables, chairs, and beer mugs. To intuitionism, the formalist claim that mathematics is formal structures is as mistaken as if one were identifying music with musical scores.

In the final part of "Intuitionism and Formalism", Brouwer concentrated on showing that when mathematical intuition has been given its proper role, then the paradoxes that had haunted mathematics will simply disappear. Intuitionism safeguards mathematics from all kinds of inconsistencies and makes metamathematical investigations superfluous.

According to intuitionism, mathematics is a product of the human mind. It is not rooted in logic, nor in language. The role of language is only to communicate features of mental constructions. Mathematical thinking cannot be captured by language, and definitely not by formal language. Mathematical thinking is not rooted in logic, but logic can be seen as an extract of mathematical thinking. Logic is dependent on mathematics. Mathematical entities do not have an independent existence; they only exist in so far as they have been constructed. Mathematical theorems concern the constructed entities, and they are true when their validity can be demonstrated through constructive proof.

7.2 Constructive Proofs

According to intuitionism, many of the proofs used in mathematics are not valid. The paradoxes that brought mathematics into a foundational crisis would evaporate if all proofs were "constructive". A constructive proof is one which brings about the relevant mathematical entities, and which constructs the truth of theorems.

What these rather vague formulations mean can be illustrated with an example. Intuitionism does not accept any form of indirect proof, which has the following format: for proving a theorem *T*, one assumes $\neg T$, and shows that this leads to a contradiction; consequently, one claims that *T* has been proved. A classic proof concerns the existence of infinitely many prime numbers. One starts assuming the negation of the theorem by stating that a maximum prime number exists. This number we can call p_{max}. The total sequence of prime numbers is then p_1, p_2, p_3, $\ldots p_{max}$. We define the number *N* by multiplying all the prime numbers and add 1:

$$N = p_1 \times p_2 \times p_3 \ldots \times p_{max} + 1$$

As with any other number, *N* can be factorised in prime numbers in a unique way. We consider one of the prime numbers in the factorisation of *N*, which we can call p_x. The prime number p_x cannot be any of the prime numbers $p_1, p_2, \ldots p_{max}$, as a prime number cannot be a factor in two consecutive numbers, namely $p_1 \times p_2 \times p_3 \ldots \times p_{max}$ and $p_1 \times p_2 \times p_3 \ldots \times p_{max} + 1$. Therefore, p_x must be different from all the prime numbers up to p_{max}, and p_x must be bigger than p_{max}. As p_{max} was assumed to be the biggest prime number, we have reached a contradiction, and we must conclude that our initial assumption was false. Therefore, infinitely many prime numbers exist.

This proof, however, cannot be considered constructive; it is indirect. By not accepting indirect proof, intuitionism abolishes all forms of Platonism in the philosophy of mathematics. According to Platonism, the set of prime numbers is either finite or infinite, and hence if the assumption that a maximum prime number exists leads to a contradiction, then one must conclude that infinitely many prime numbers exist. Indirect proof is in accordance with Platonism assuming the existence of a mathematical reality independent of human beings.

By not accepting indirect proof, intuitionism embraces a logic different from the one normally assumed to be valid in mathematics. Different from the logic presented by Gottlob Frege (1983) in the *Begriffsschrift*, from the one presented by Alfred Whitehead and Bertrand Russell (1910, 1913a, b) in *Principia Mathematica*, and from the logic assumed by formalism. While classical logic holds $p \vee \neg p$ to be a valid logical proposition, intuitionism does not. The formula $p \vee \neg p$ expresses the principle of excluded middle, according to which any statement *p* must either be true or false. The validity of the excluded middle is the basis for indirect proof. The principle is also referred to as the principle of the excluded third, or as the *Tertium non datur*. According to intuitionism, one can be in a situation where, for a given mathematical formula *F*, neither F nor $\neg F$ has been proven.

In his *Cambridge Lectures*, Brouwer (1981) made the following observation with respect to the principle of the excluded third:

> The belief in the universal validity of the principle of the excluded third in mathematics is considered by the intuitionists as a phenomenon of the history of civilization of the same kind as the former belief in the rationality of π, or in the rotation of the firmament about the Earth. (p. 7)

According to Brouwer the principle of the excluded middle could be considered a historically rooted conviction similar to many other dubious convictions. The assumed validly of $p \vee \neg p$ is nothing but a cultural phenomenon, which can be explained along the same lines as many other cases of cultural convictions. Brouwer did acknowledge that in some particular cases the principle of the excluded middle does work: it could be in everyday situations; it could also be in some more particular mathematical cases. But as a general principle to be used in mathematics, it is invalid.[4]

Indirect proof has been broadly applied in mathematics, so it seems that the mathematical edifice suffers a devastating earthquake caused by intuitionism. However, according to intuitionism, some reconstruction is possible. In several cases there might exist an alternative constructive proof to an indirect proof. Much work in intuitionistic mathematics has been dedicated to rebuilding part of mathematics by showing how one can provide constructive proofs for theorems that have previously only been demonstrated through indirect proof.

As an illustration of what this could mean, let us look again at the theorem that infinitely many prime numbers exist. The proof we presented was indirect, but it can easily be reformulated as a constructive proof. First, we reformulate the theorem as: *for any prime number* p_n, *it is always possible to construct a prime number* p_{n+1} *that is bigger than* p_n. We define again the number N as $N = p_1 \times p_2 \times p_3 \ldots \times p_n + 1$, and we consider one of the prime numbers p_x in the factorisation of N. Again we observe that this prime number p_x cannot be any of the prime numbers $p_1, p_2, \ldots p_n$, and we conclude that p_x must be bigger than p_n. As we in fact can determine p_x by a factorisation of N, we have shown how, given a prime number p_n, we are able to construct a prime number p_x bigger than p_n. So, from the perspective of intuitionism there also exist infinitely many prime numbers.

To transfer the indirect proof of the existence of infinitely many prime numbers into a constructed proof was not any big deal; it was just a case of rephrasing the proof. Many mathematical theorems can be "saved" by producing a constructive proof, but doing so could be a laborious task. The proof that Brouwer (1911) originally gave for the fixed-point theorem was not constructive, but it has been reworked into a constructive format.[5] So the fixed-point theorem is also a theorem in intuitionistic mathematics.

In the hands of intuitionism, there is no space for any hint of relativism. Mathematics intuition as *Formen der Anschauung* is independent of personal experiences,

---

[4] For a discussion of the principle of excluded middle, see, for instance Kolmogorov (1967).

[5] See Kellogg, Li and Yorke (1976).

social structures, and historical developments. Principally, we are dealing with an *a priori* absolute. *Formen der Anschauung* cannot be any different, and so mathematics cannot be any different either. Intuitionism advocates for absolutism in mathematics, albeit a different form of absolutism than both logicist absolutism and formalist absolutism. Intuitionism advocates a mental absolutism: mathematics is the same for all who understand![6]

7.3 Intensional Logic

Making the truth tables for the following two formulas will show that they are tautologies:

$$p \Rightarrow (\neg p \Rightarrow \mathrm{q})$$

$$(p \Rightarrow p) \vee (\mathrm{q} \Rightarrow p)$$

They are theorems in classic axiomatisations of propositional calculus, but according to intuitionism they are not valid.

The first theorem states that if p is true, then $\neg p$ will imply q, whatever q might be. A false statement will imply whatever: if the moon was made of blue cheese, then Iceland would be flying. This is valid logical reasoning according to the formula $p \Rightarrow (\neg p \Rightarrow \mathrm{q})$. The second formula states that for any two statements p and q, either p will imply q, or q will imply p. The two statements could be "2 + 2 = 4" and "infinitely many prime numbers exist". According to the formula we will have either that "2 + 2 = 4" implies "infinitely many prime numbers exist", or "infinitely many prime numbers exist" implies "2 + 2 = 4".

At times, classic propositional logic may not resonate with what is considered sound mathematical reasoning. One particular problem is related to the concept of implication:

$$p \Rightarrow \mathrm{q}$$

According to classic logic, the meaning to be associated to the symbol "⇒" can be completely specified by means of a truth table. But the consequence is that one comes to operate with a notion of implication that is far removed from a transparent intuitive interpretation. It seems, in accordance with common sense, that if one claims that p implies q, then one should be ready to present a deductive route leading from p to q. This in fact appears always to take place in mathematical proof. Through

[6]A captivating introduction to intuitionism is found in Heyting (1971). Studies of intuitionism can be found in Dummett (1977) and Troelstra (2011). Recent contributions to intuitionistic and constructive mathematics can be found in Bishop (1967), who provides a foundation of constructive analysis; and Lorenzen (1971), who gives a constructive introduction to classical analysis.

a proof one is supposed to show that if p is the case, then q *must* also be the case. The experience of logical necessity is an essential part of logical deduction. However, as long as we interpret implication in terms of truth tables, the idea of necessity disappears.

Intuitionistic logic finds that mathematical deduction should establish necessary links between mathematical statements. A huge step in clarifying the nature of intuitionistic logic was taken in 1930, when Arend Heyting (1898–1980) published the paper "Die formalen Regeln der intuitionistischen Logik" ("The formal rules of intuitionistic logic"). Here, Heyting (1930) presented intuitionistic logic in the form of an axiomatic formal system. Among its theorems one does not find the formula $p \vee \neg p$, nor the two formulas that we have just considered: $p \Rightarrow (\neg p \Rightarrow \mathrm{q})$, and $(p \Rightarrow p) \vee (\mathrm{q} \Rightarrow p)$. Through his formalisation Heyting tried to express some of the basic properties of constructive mathematical deduction.

While logicism considers logic to be the foundation of mathematics, intuitionism sees the relationship between mathematics and logic quite differently. It positions mathematics as a unique intuitive process that cannot be reduced to anything else. This process is steered by its own dynamics. Intuitionistic logic is an extract from this dynamic. Brouwer, however, was not pleased with Heyting's formalisation, as he did not find that the mental acts of mathematics could be captured by any formal system. In spite of Brouwer's reservations, Heyting's formalisation became an important step in the further development of logic.

Classic logic and intuitionistic logic are two different kinds of logic. In classic logic the meanings of the logical connectives can be clarified in term of truth tables, while in intuitionistic logic they cannot. Classic logic concentrates on the extensions of concepts (which are sets) and the extension of statements (which are truth values), and consequently it is referred to as *extensional logic*. Intuitionistic logic focuses on the intensions of concepts and statements, and consequently it is referred to as *intensional logic*. In his theory of meaning, Frege drew a distinction between sense and reference. While extensional logic concentrates on "references", intensional logic pays particular attention to "senses".

In Heyting's formalisation one finds the symbol $\Rightarrow$, and as in classic logic this also represents "implication". However, the intuitionist notion of implication is radically different from the notion of implication in classic logic. With his formalisation of intuitionistic logic, Heyting led the way for a better grasp of the nature of necessity associated with the processes of constructive deduction.[7]

7.4 Inspirations

Brouwer was dedicated to disseminating the intuitionist views on mathematics, and he inspired and attracted many people.

[7] For a discussion of the philosophical basis of intuitionistic logic, see Dummett (1983).

In 1927, Brouwer lectured in Berlin, and among the audience was Hans Freudenthal (1905–1990). He was among the up-and-coming young stars in mathematics; his topic was topology, and in 1930 he defended his doctoral thesis within this field.[8] Brouwer appreciated the young Freudenthal, both for his mathematical insight and for his broader philosophic outlook. Freudenthal also demonstrated a deep understanding of the intuitionist approach to mathematics. In Berlin, Karl Löwner had given a course in calculus, where he had carefully elaborated all proofs according to intuitionistic standards. Freudenthal had followed this course.

In 1930 Brouwer invited Freudenthal to come to Amsterdam as his research assistant. Freudenthal, however, was reluctant to accept a job in Amsterdam which meant that he would have to leave Berlin. At that time Göttingen and Berlin were the two most influential centres for mathematics research, not only in Germany but in the entire world. As Freudenthal put it: moving to Amsterdam was like moving to a modest German university.

Brouwer did not live in Amsterdam, but in Laren, a small city close by. He insisted that his assistants should live in Laren as well, as he might need to talk with them urgently. A life in a small Dutch city was not in accordance with Freudenthal's taste. He appreciated big cities, and Amsterdam seemed a tolerable substitute for Berlin. But definitely not Laren. Brouwer insisted, and so did Freudenthal. In the end Brouwer gave in, although on the condition that Freudenthal got a telephone installed so that Brouwer could call him whenever he wanted. Freudenthal got the telephone, but Brouwer never called.

Freudenthal was of Jewish origin, and during the Second World War when the Netherlands was occupied by Germany, he lost his job at the university. He was put in a concentration camp, but managed to escape, and until the Netherlands was liberated, he remained in hiding. When the war ended, it was generally assumed that Freudenthal would be given back his position at the University of Amsterdam, but this did not happen. The relationship between Brouwer and Freudenthal had been ambiguous, and after the war Brouwer did not show any enthusiasm for having Freudenthal back as a colleague. Instead Freudenthal obtained a post at the University of Utrecht, and here he developed a huge interest in mathematics education. Freudenthal's view on mathematics education was inspired by the intuitionist conception of mathematics, although this is seldom explicitly acknowledged.

In 1928 Brouwer gave two lectures in Vienna, the first having the title "Mathematics, Science and Language". At that time Ludwig Wittgenstein had abandoned philosophy; he had worked as primary school teacher, and also as an architect in designing a house for one of his sisters. Some contact between Wittgenstein and the Vienna Circle, however, had continued, and people from the circle had urged Wittgenstein to attend Brouwer's lecture.

This became an event of extreme importance for Wittgenstein and for philosophy in general. The lecture kindled Wittgenstein's philosophic devotion, and it opened up a new direction for his thinking—a direction away from the ideas that he had

[8] See Freudenthal (1931). Today some mathematical theorems bear his name.

presented in the *Tractatus Logico-Philosophicus*. As was common, people from the Vienna Circle met after the lecture at a coffee house to enjoy each other's company and have discussions. Herbert Feigl from the circle describes Wittgenstein's reaction in the following way: "He became extremely voluble and began sketching ideas that were the beginning of his later writings. That evening marked the return of Wittgenstein to strong philosophical interest and activities" (Feigl, quoted after Dalen, 2013, p. 517).

What could have inspired Wittgenstein so profoundly? It might be some of Brouwer's observations with respect to the role of language. In the *Tractatus*, Wittgenstein sided with, among others, Bertrand Russell, in claiming that natural language is inappropriate for dealing with scientific and philosophical issues. Formal language is necessary. In the *Tractatus* Wittgenstein talks about *the* language, having in mind the formal language presented by Whitehead and Russell (1910, 1913a, b) in *Principia Mathematica*, and he insisted that this is the only appropriate language for science. Brouwer, however, did not share any such veneration for formalisations. He saw mathematics as mental acts, which cannot be captured by any formal language. For him, language existed in a variety of forms that may play different roles, including expressing facets of the mental acts that constitute mathematics. Such ideas might have helped Wittgenstein to abandon the dogma that a formal language is *the* language, and to bring him on a route directed towards the concept of "language games".

7.5 Limitations and Impacts

Logicism, in particular the form presented by Frege, assumes a mathematical realism according to which mathematical entities have an existence of their own, independent of our knowledge about them. According to logicism, mathematical reality is the world of references. This ontological realism is the foothold of logicist absolutism. According to formalism, mathematics can be developed without any ontological commitments. Mathematical entities do not need any specification; the only thing that is relevant is the stipulated relationship between such entities. This gives formalist absolutism an if-then format. Intuitionism does not accept realism, and nor does it assume the ontological indifference associated with formalism. According to intuitionism, mathematical entities exist by being constructed. There is, however, only one way of creating such constructions, and this ensures the intuitionist version of absolutism.

As with logicist and formalist absolutism, intuitionist absolutism amputates any social issues from the philosophy of mathematics. The fixation on ensuring that mathematical truths are absolute creates an intrinsic focusing, which immerses mathematics into an ethical vacuum. Mathematics appears to be an absolute, separated from social affairs. Mathematics can be deified as a pure and unique epistemic prodigy. Whilst coming from their own different starting points, the three philosophies—logicism, formalism, and intuitionism—eventually arrive together to each deify mathematics, seeing it as riding above everyday concerns. This is the reason that I have referred to logicism, formalism, and intuitionism as the Three Magi.

Intuitionism has had a huge impact on the further development of logic. Intuitionistic logic is part of the extended family of intensional logic. One of them is modal logic, which operates with the notions of "necessity" and "possibility". In a very short paper, hardly longer than one page, Kurt Gödel (1933) shows that intuitionistic logic is equivalent to a version of modal logic. This means in particular that the intuitionist conception of $p \Rightarrow$ q can become interpreted as "p implies q by necessity". This captures the intuitionist understanding of mathematical proof; it also captures the common-sense understanding of proof. In fact, it seems to capture the idea of proof that is common in mathematical practice. The intuitionist understanding of proof has had a huge impact on the development of computer-based mathematical proof and on the development of artificial intelligence.[9]

The intuitionist concept of mathematics has opened new views on mathematics education. Such views were first of all formulated by Freudenthal. That he, after the Second World War, had to accept a job at the University of Utrecht, was felt by him to be a severe degradation. However, staying in Utrecht opened up new opportunities for him. He became engaged in mathematics education, and from 1967 to 1970 he was president of the International Commission on Mathematical Instruction (whose first president was Felix Klein). In 1971 he went on to found the Institute for the Development of Mathematical Education (IOWO) at Utrecht University.

While the Modern Mathematics Movement assumed a structuralist conception of mathematics, Freudenthal was inspired by the intuitionist view of mathematics. He never did become an ardent intuitionist as, for instance, Heyting did, but, like Heyting, he was familiar with the intuitionist approach to mathematics.[10] In his educational works Freudenthal only included a few references to Brouwer, but in the preface to *Mathematics as an Educational Task*, he mentioned that he was inspired by Brouwer's conception of mathematics.[11] Freudenthal's confrontation of the Modern Mathematics Movement was similar to Brouwer's clash with formalism. Both Freudenthal and Brouwer argued that formal structures did not capture the essence of mathematical thinking. While Brouwer highlighted that mathematics is a *mental* activity, Freudenthal repeated many times that mathematics is a *human* activity.[12] They first of all saw mathematics as an activity and not as a structure. While the Modern Mathematics Movement tried to teach students mathematical scores, Freudenthal tried to engage them in playing the music of mathematics.

---

[9]For a discussion of intuitionism and computing, see for instance Martin-Löf (1968, 1982).

[10]Brouwer's collected works appeared in two volumes. The first, *Collected Works, Vol 1: Philosophy and Foundations of Mathematics* is edited by Heyting, while the second, *Collected Works, Vol. 2: Geometry, Analysis, Topology and Mechanics* is edited by Freudenthal.

[11]See Freudenthal, 1973, p. ix). In the following I draw on Skovsmose (in progress), where I have discussed more carefully the relationship between Brouwer's intuitionism and Freudenthal's educational ideas.

[12]See, for instance, Gravemeijer and Terwel (2000).

Logicism operates with a theory of meaning condensed in Frege's distinction between *Sinn* and *Bedeutung*, between sense and reference. According to Frege, mathematics must fully concentrate on references to concepts and statements. Brouwer rejected such a claim, and Freudenthal's conception of mathematics was guided by a similar rejection. According to Freudenthal (1973), mathematics is a human activity taking place in contexts "fraught with relations" (p. 78), and consequently the meaning of an activity becomes formed through a complexity of relationships. This broad understanding of meaning became an important component of the approach to mathematics education that Freudenthal advocated and which IOWO developed into many practical activities.[13] Freudenthal's interpretation of "fraught with relations" is, however, to a great extent depoliticised; it does not refer to socio-political relationships. Freudenthal confronted the structuralist conception of mathematics, but not the modern image of mathematics as being detached from political controversies.

Although the modern philosophies of mathematics—logicism, formalism, and intuitionism—have had a profound impact on mathematical practices in research, application, and education, they have remained oblivious to any such impacts. They stay fully focused on logical and mathematical specificities in order to clarify the nature of mathematical theories and truths. Never before has the philosophy of mathematics expressed itself with such a degree of technical sophistication. The modern philosophies of mathematics remain oblivious to any socio-political issues related to mathematics. They treat mathematics as an isolated piece of knowledge that could be investigated unconnected to other forms of knowledge as well as to any socio-political issues. In this way the modern philosophies of mathematics contribute to a deceitful image of mathematics as being a unique human achievement that deserve to be glorified.

During Modernity, mathematics and science demonstrated an extreme degree of male dominance; this dominance has been repeated by the modern philosophies of mathematics. Apparently, all major contributions to these philosophies come from men. This male dominance in the philosophy of mathematics has continued. *The Oxford Handbook of Philosophy of Mathematics and Logic* (Shapiro, 2005) includes 26 contributions, of which three are written by women. In trying to grasp the possible impact of this bias, one can consider what a feminist epistemology of mathematics might include (Burton, 1995), and how feminist theories might inform mathematical practices (McGuire, 2020).[14] In spite of these biases, the Modern philosophies of mathematics broadcast the dogma that mathematics is objective, neutral, and not serving any particular political or social interests.

---

[13] See Five Years IOWO (1976). In Skovsmose (in progress), I present an example from this book. See also Freudenthal (1978, 1983, 1991).

[14] For a discussion of gender, mathematics, and mathematics education, see Kollosche et al. (in progress).

7.6 Expelled from Wonderland

Initially Hilbert had been impressed by Brouwer's work in topology, and was ready to bring him to Göttingen. However, as Brouwer formulated the intuitionist understanding of mathematics as a confrontation with formalism, Hilbert came to a different view on Brouwer. Hilbert could not accept that mathematics should be under the restrictions set by intuitionism. He wanted, for instance, to apply the principle of excluded middle, the *Tertium non datur*, in mathematical proof. In "The Foundations of Mathematics", based on a lecture Hilbert gave in 1927, he outlined his fundamental ideas of the proof theory, by means of which he wanted to save mathematics from paradoxes and uncertainties. Here, Hilbert (1967) stated:

> Taking the principle of excluded middle from the mathematician would be the same, say, as proscribing the telescope to the astronomer or to the boxer the use of his fists. To prohibit [...] the principle of the excluded middle is tantamount to relinquishing the science of mathematics altogether. (p. 476).

Hilbert wanted to maintain all the mathematical richness established by indirect proof. He wanted to preserve Cantor's Wonderland, which Hilbert had also referred to as Cantor's paradise (see Hilbert, 1983, p. 191). In the end, Hilbert saw Brouwer's intuitionism as being destructive for mathematics.

The abolishing of the principle of the excluded middle would destroy Cantor's paradise. The series of transfinite cardinal numbers was generated by the observation that, for any infinite set A, the cardinality of the powerset of A will be greater than the cardinality of A: we refer to the powerset of A as $\mathcal{P}(A)$, and Cantor showed that:

$$Card(A) < Card(\mathcal{P}(\mathrm{A}))$$

His proof was indirect.[15] He assumed that a surjective function from A to $\mathcal{P}(A)$ exists, and showed that this assumption leads to a contradiction. As a consequence, he concluded that no such surjective function does exist, meaning that the cardinality of $\mathcal{P}(A)$ must be greater that the cardinality of A.

By considering Cantor's proof to be invalid, intuitionism deactivated the formidable generator of transfinite cardinal numbers. Brouwer (1983) pointed out that "the intuitionist recognizes only the existence of denumerable sets" (p. 81). With the term "denumerable sets" he referred to sets that are equipotent with the set of natural numbers or with subsets of the set of natural numbers. Brouwer simply closed down Cantor's Wonderland.

[15] The proof is outlined in the Sect. 6.5.

References

Benacerraf, P., & Putnam, H. (Eds.). (1983). *Philosophy of mathematics: Selected readings* (2nd ed.). Cambridge University Press.

Bishop, E. (1967). *Foundations of constructive analysis*. McGraw-Hill.

Brouwer, L. E. J. (1911). Über Abbildungen von Mannigfaltigkeiten. *Mathematische Annalen, 71*, 97–115.

Brouwer, L. E. J. (1981). In D. van Dalen (Ed.), *Brouwer's Cambridge lectures on intuitionism*. Cambridge University Press.

Brouwer, L. E. J. (1983). Intuitionism and formalism. In P. Benacerraf & H. Putnam (Eds.), *Philosophy of mathematics: Selected readings* (2nd ed., pp. 77–89). Cambridge University Press.

Brouwer, L. E. J. (1996). Life, art, and mysticism. *Notre Dame Journal of Formal Logic, 37*(3), 389–429.

Burton, L. (1995). Moving towards a feminist epistemology of mathematics. *Educational Studies in Mathematics, 28*, 275–291.

Dalen, D. V. (2013). *L. E. J. Brouwer—Topologist, intuitionist, philosopher: How mathematics is rooted in life*. Springer.

Dummett, M. (1977). *Elements of intuitionism*. Clarendon Press.

Dummett, M. (1983). The philosophical basis of intuitionistic logic. In P. Benacerraf & H. Putnam (Eds.), *Philosophy of mathematics: Selected readings* (2nd ed., pp. 98–129). Cambridge University Press.

Five Years IOWO. (1976). *Educational Studies in Mathematics, 7*(3).

Frege, G. (1983). Begriffsschrift: A formula language, modelled upon that of arithmetic, for pure thought. In J. van Heijenoort (Ed.), *From Frege to Gödel: A source book in mathematical logic, 1879–1931* (2nd ed., pp. 1–82). Cambridge University Press.

Freudenthal, H. (1931). Über die Enden topologischer Räume und Gruppen. *Mathematische Zeitschrift, 33*, 692–713.

Freudenthal, H. (1973). *Mathematics as an educational task*. Reidel.

Freudenthal, H. (1978). *Weeding and sowing*. Reidel.

Freudenthal, H. (1983). *Didactical phenomenology of mathematical structures*. Reidel.

Freudenthal, H. (1991). *Revisiting mathematics education: China lectures*. Kluwer.

Gödel, K. (1933). Eine Interpretation des intuitionischen Aussagenkalküls. *Ergebnisse Math. Colloq, 4*, 39–40.

Gravemeijer, K. J., & Terwel, J. (2000). Hans Freudenthal: A mathematician on didactics and curriculum theory. *Journal of Curriculum Studies, 32*(6), 777–796.

Heyting, A. (1930). *Die formalen Regeln der intuitionistischen Logik*. Akademie der Wissenschaften. phys.-math, 42–56, 57–71, 158–169.

Heyting, A. (1971). *Intuitionism: An introduction*. North Holland Publishing.

Hilbert, D. (1967). The foundations of mathematics. In J. van Heijenoort (Ed.), *From Frege to Gödel: A source book in mathematical logic, 1879–1931* (pp. 464–479). Harvard University Press.

Hilbert, D. (1983). On the infinity. In P. Benacerraf & H. Putnam (Eds.), *Philosophy of mathematics: Selected readings* (2nd ed., pp. 183–201). Cambridge University Press.

Kant, I. (1973). *Critique of pure reason*. (Norman Kemp Smith, Trans.). The MacMillan Press.

Kellogg, R. B., Li, T. Y., & Yorke, J. A. (1976). A constructive proof of the Brouwer fixed-point theorem and computational results. *SIAM Journal on Numerical Analysis, 13*(4), 473–483.

Kollosche, D., Steflitsch, D., & Deweis-Weidlinger, K. M. (in progress). Gender, mathematics, and mathematics education. In B. Greer, D. Kollosche, & O. Skovsmose (Eds.), *Breaking images: Iconoclastic analyses of mathematics and its education*. Open Book Publishers.

Kolmogorov, A. N. (1967). On the principle of excluded middle. In J. van Heijenoort (Ed.), *From Frege to Gödel: A source book in mathematical logic, 1879–1931* (pp. 414–439). Harvard University Press.

Locke, J. (1997). *An essay concerning human understanding*. Penguin Books.
Lorenzen, P. (1971). *Differential and integral: A constructive introduction to classical analysis*. University of Texas Press.
Martin-Löf, P. (1968). *Notes on constructive analysis*. Almquist and Wiksell.
Martin-Löf, P. (1982). Constructive mathematics and computer programming. In L. J. Cohen, J. Los, H. Pfeiffer, & K.-P. Podewski (Eds.), *Logic, methodology and philosophy of sciences VI*. North-Holland Publishing Company.
McGuire, L. (2020). Feminist theories informing mathematical practice. In B. Sriraman (Ed.), *Handbook of the mathematics of the arts and sciences* (pp. 1–23). Springer.
Pambuccian, V. (2000). Brouwer's intuitionism. In B. Sriraman (Ed.), *Handbook of the history and philosophy of mathematical practice*. Springer.
Ravn, O., & Skovsmose, O. (2019). *Connecting humans to equations: A reinterpretation of the philosophy of mathematics*. Springer.
Shapiro, S. (Ed.). (2005). *The Oxford handbook of philosophy of mathematics and logic*. Oxford University Press.
Skovsmose, O. (in progress). Intuition revived. In B. Greer, D. Kollosche, & O. Skovsmose (Eds.), *Breaking images: Iconoclastic analyses of mathematics and its education*. Open Book Publishers.
Troelstra, A. S. (2011). History of constructivism in the 20th century. In J. Kennedy & R. Kossak (Eds.), *Set theory, arithmetic, and foundations of mathematics: Theorems, philosophies* (Lecture notes in logic, Vol. 36) (pp. 150–179). Cambridge University Press.
Whitehead, A., & Russell, B. (1910/1913a/1913b). *Principia mathematica I-III*. Cambridge University Press.

Chapter 8
Mathematics as Mathematicians' Practice

Abstract The philosophy of mathematical practice attempts to give the philosophy of mathematics a new beginning. Essential for this beginning is to acknowledge that mathematics is constructed by humans. Hence, as with any other kind of human knowledge construction, mathematics is fallible, corrigible, tentative, and evolving. According to the philosophy of mathematical practice, mathematics can be defined simply as the mathematicians' practice—it is what mathematicians do. For a philosophy of mathematical practice, it is important to address features of this practice instead of being preoccupied with problems defined through philosophic traditions. It is important to explore the concept of proof, not by embracing any specific ideals, but by considering the variety of ways proof is accomplished in practice. As with logicism, formalism, and intuitionism, so also the philosophy of mathematic practice operates in an ethical vacuum by not considering what might be the social impact of bringing mathematics into action.

Keywords Philosophy of mathematical practice · Logicism · Formalism · Visualisation · Computer-based proof · Diagram-based proof · Absolutism · Ethical vacuum · Wonderland

So, what is mathematics? Mathematics is logic! answered the logicist. Mathematics is formal structures! answered the formalist. Mathematics is mental acts! answered the intuitionist. The answer that I am going to address in this chapter is: mathematics is the mathematicians' practice! This might sound a trivial answer. It might sound circular. It might sound like a joke.

But there is much more to this answer. We currently live in São Paulo, and imagine that we receive some visitors from Far Away, who have never heard about football. They ask us: What is this football? Since we want to give the visitors a precise answer, we look up the official FIFA rules and start explaining them one by one. We soon realise that this approach does not work, but why not take our visitors to a real football match? They like the idea, so we buy tickets to a match between Corinthians and Palmeiras. These two teams are both based in São Paulo and play in

O. Skovsmose, *Critical Philosophy of Mathematics*, Advances in Mathematics Education, https://doi.org/10.1007/978-3-031-71375-0_8

the Serie A.[1] The match is important not only for the teams' position in the league table, but also since it is part of the city rivalry. We can expect it to be a hard-fought match.

The stadium is huge, with more than 40,000 spectators. Some are wearing black T-shirts while others are dressed in green T-shirts. When we enter the stadium, we feel swallowed up by the loud noise from the spectators. When the referee and the players come on to the field, we explain to our visitors that a football match is between two teams, and that the referee does not belong to either of them. Yes, two teams; there cannot be three teams in a football match. All the players in each team are dressed in shirts of the same colour. We also need to explain that many of the spectators show their allegiance by dressing in the colours of the team they support, but that they are not allowed to take part in the game. Our visitors from Far Away ask whether the match has started. It has not, but we tell them when it begins. The way the players are behaving seems strange to them. We explain that, as it is football, the players are only allowed to kick the ball. They are not allowed to touch the ball with their hands. But we then have to explain that there is one exception to this—the goalkeeper can use his hands and pick up the ball. The visitors from Far Away are concerned about the gender bias: Why are all the players men? We tell them a bit about the history of football, that it was a sport for men but there are now tournaments for women—although no mixed teams. In tennis there are mixed doubles (the visitor knows about tennis), so why no mixing in football? Before we can provide an answer, a player picks up the ball with his hands. But this is not allowed, our visitors say! We have to explain that this is because the ball has been kicked out of the field, and that the player then is allowed to pick it up and make a throw-in. After some time, the whole stadium explodes in a ferocious roar which frightens our visitors. What is happening? Gooooooal! Gooooooal! Gooooooal! It is a goal, we explain. When the roar has faded away, we explain about the importance of scoring goals. The visitors from Far Away are disappointed when all the players suddenly leave the field. Is it the end of the game? No, no, this is just half-time. The players are just taking a break. Eventually, they come back and continue the game. And more goals are scored until a whistle was blown and half of the crowd cheers.

Our visitors from Far Away enjoyed watching the football match, and thanked us very much for bringing them to the stadium. Yes, this was a real match! Having experienced it, they came to know much about football. In this way, we defined football as the football players' practice.

[1] Serie A is similar to the Bundesliga in Germany, La Liga in Spain, and the Premier League in England.

8.1 A New Beginning

In 1979, Reuben Hersh (1927–2020) published the paper "Some Proposals for Reviving the Philosophy of Mathematics". This publication created a shift in the philosophy of mathematics. It abandoned the agenda set by logicism, formalism, and intuitionism, and cleared the way for a new approach that is now referred to as the *philosophy of mathematical practice*.

Hersh (1979) found that the philosophy of mathematics had lost contact with what was taking place in mathematical practice, which he first of all saw as *the thing that mathematicians do*—mathematicians' practice. The philosophy of mathematics has come to focus on problems that might have a philosophical significance, but appear irrelevant for the working mathematician. Initially, the agenda of logicism, formalism, and intuitionism was set by mathematicians, but subsequently the agenda was hijacked by philosophers. Hersh was deeply concerned about the philosophic inbreeding that had taken place in the philosophy of mathematics.

One might find an example of what Hersh had in mind by studying the *Philosophy of Mathematics*, written by Stephan Körner (1968), professor of philosophy at the University of Bristol. In this book Körner crafted a careful exposition of logicism, formalism, and intuition, and then raised some critical issues with respect to each of the three positions. He found, for instance, that logicism creates a conflation of empirical and non-empirical concepts; that formalism does not seem to recognise the ample disconnection between mathematics and perception; and that intuitionism is vulnerable to conflicting reports on experienced self-evidence. It is not difficult to see that Hersh, as a mathematician, could not feel at home among such challenges. The philosophical inbreeding continued after Hersh published his article. In *Thinking About Mathematics: The Philosophy of Mathematics*, Stewart Shapiro (2000), at the time professor of philosophy at the Ohio State University, investigated the distinction between *ante rem realism* and *in re realism*. The first kind of realism, Shapiro associates with Plato, while he associates the second kind with Aristotle. It is extremely difficult to associate such a distinction with any experiences of the working mathematician.[2]

Hersh wanted the philosophy of mathematics to be guided by an agenda experienced as relevant by working mathematicians. He found, however, that the concept of mathematics that is often held by the working mathematicians themselves is not congruent with the mathematicians' own experiences when doing mathematics. Very elegantly Hersh (1979) captured this incongruence by pointing out that, typically, a working mathematician is "a Platonist on weekdays and a formalist on Sundays" (p. 32). When doing research, mathematicians might feel that they are trying to discover mathematical truths; while being questioned about what they are doing and why, they might hide behind a formalist discourse, emphasising that mathematics is a formal game. Hersh did not want to assume Platonism nor

[2]An updated expression of the inbreeding that Hersh is concerned about can be found in Shapiro (2005).

formalism; he searched for something different. He wanted the philosophy of mathematics to abandon the search for foundations and stop dreaming about establishing a regime of absolute truths. To Hersh it was important to move away from the agenda set by the Three Magi. He highlighted that, although many of the philosophical problems formulated by logicism, formalism, and intuitionism might have captured philosophers' attention, we are free to discard them "if we choose to do so" (p. 31). Many of the classic problems related to the foundation of mathematics and to the understanding of the nature of mathematical truths are not essential to mathematicians' practice. Such problems would simply evaporate as part of a new beginning of a philosophy of mathematics.

In *Proofs and Refutations* by Imre Lakatos (1976), Hersh found inspiration for a new beginning. Lakatos considered it aimless to search for mathematical certainties; instead, he found it important to recognise that mathematics is fallible. Hersh (1979) agreed, and he found mathematical knowledge to be "fallible, corrigible, tentative and evolving, as is every other kind of human knowledge" (p. 43). This observation was central to Hersh's understanding of mathematics. Mathematicians' practice creates results that are fallible; they might need corrections, they are tentative, they are part of an ongoing development without any termination in sight. Hersh's constructivism was radically different from the one advocated by intuitionism. While intuitionism considered constructions to be of a mental format, Hersh saw mathematical constructions as part of a real-life mathematical practice, which itself is a social practice.

Hersh did not search for a sophisticated answer to the question: What is mathematics? For him there is no profound answer to this question waiting to be excavated through meticulous philosophical investigations. The question can be answered in a straightforward way: mathematics is constituted through mathematicians' practice.

Hersh found mathematical entities to be "invented or created by humans" (p. 44). However, we are not dealing with arbitrary creations, since mathematical constructions are elaborations, extensions, modifications, or reconstructions of already existing constructions. New mathematical objects "arise from activity with already existing mathematical objects" (p. 44). Once created, "mathematical objects have properties which are well-determined, which we may have great difficulty in discovering, but which are possessed independently of our knowledge of them" (p. 44). Mathematical objects are created, and when created they come to form part of mathematical realty. One can make discoveries of properties of such constructed realities. They have "well-determined properties of their own, which we may or may not be able to discover" (p. 45).

The philosophy of mathematics that Hersh wanted to establish as a new beginning saw mathematics as a human construction, constantly in development. Mathematical theorising is temporary; there are no eternal truths to be found in mathematics. Mathematical development is propelled forward by the mathematicians' practice.

8.2 Proofs in Practice

Proving is a defining characteristic of mathematical practice. The Three Magi all provided detailed descriptions of what can be accepted a valid mathematical proof. Logicism presented an axiomatisation of logic, which theorems should include all the forms of mathematical reasoning that could be applied in mathematical proof. Formalism described mathematical proof with reference to a formalised axiomatic system as a sequence of formulas $A_1, A_2, \ldots, A_n$, where any formula A_i $(1 \leq i \leq n)$ is either an axiom of the system, or can be derived from one or more of the formulas in the sequence $A_1, \ldots, A_{i-1}$. In a sequence such as $A_1, A_2, \ldots, A_n$, the last formula, $A_{n,}$ can be called a theorem. Intuitionism found that mathematical proofs need to be constructive, which restricts the scope of mathematical reasoning to being only that which is captured, as least to some degree, by intuitionistic logic.

Such specifications of what counts as a mathematical proof are not, however, a defining feature of a philosophy of mathematical practice. This philosophy does not try to present any *a priori* definition of what counts as a mathematical proof, but tries instead to conceptualise what is taking place in the mathematical practice. This practice includes many different ways of proving, and the philosophy of mathematical practice adopts a principle of tolerance. Hersh (1979) highlighted that proofs "are not universal; they differ from one branch of mathematics to another, and from one historical epoch to another" (p. 47).

In 1979, the same year that Hersh published "Some Proposals for Reviving the Philosophy of Mathematics", Thomas Tymoczko (1943–1996) published "The Four-Color Problem and Its Philosophical Significance" (Tymoczko, 1979). For a long time the four-colour theorem has been considered a mathematical challenge. The theorem concerns the colouring of maps, where a sound principle is that countries next to each other should be coloured differently. That provides us with a practical question: How many different colours would it be necessary to use? One could have in mind a particular map, but one could also think of maps in general: How many colours are needed for making sure that countries next to each other are coloured differently, whatever map we are dealing with?

Based on trial-and-error methods it was observed that four colours might be sufficient. This conjecture was proposed in 1852, when Francis Guthrie, coming from South Africa to study in England, noticed that for a map showing the British Empire, only four different colours seemed necessary. He communicated his observation to the mathematician Augustus De Morgan, professor at London University (now University College London), and since then the four-colour conjecture has been a mathematical challenge.

In 1976 Kenneth Appel and Wolfgang Haken (1978) published a proof showing that Guthrie's guess was correct: only four different colours are necessary whichever map we are dealing with.[3] However, their approach was new: they used a computer to complete their proof. It was the first time that a computer had been assigned a

[3] See also Appel and Haken (1978) and McEvoy (2020).

principal role in mathematical proof. They did the proving by induction. With respect to natural numbers, a proof by induction includes three steps: the first one proves that a certain property applies to the number 1; the second one proves that if the property applies to a number n, then is also applies to the number $n + 1$; and the third one concludes that the property applies to all natural numbers. For completing a proof by induction, it might at times be necessary to go through different situations separately, and this was the case with respect to the four-colour theorem. However, in this case it was necessary to consider more than 1000 different situations, and it became necessary to use a computer in order to check through all these different situations.

But what kind of proof was this? It did not fit into any of the categories that have previously been cultivated in mathematics, nor did it comply with any of the criteria for proof that had been formulated by logicism, formalism, and intuitionism. Could a mathematical proof that nobody has seen exist? It appears to be a defining feature of a mathematical proof that it is public, and that it is possible to conduct a systematic step-by-step check of it. In mathematics there is nothing called a hidden proof. However, if a computer operates like a "black box", what then are we to think of a computer-based proof?

Tymoczko (1979) introduced a new agenda on the discussion of mathematical proofs. He operated with three different questions with respect to a mathematical proof: Is it convincing? Can it be formalised? Is it surveyable?

It makes sense to claim that a proof needs to be convincing; unconvincing proofs do not function in practice. "Convincing" appears to be a psychological notion, and it has not been considered by any of the classic positions in the philosophy of mathematics; on the contrary, these philosophies have tried to remove any trace of psychologism from mathematical concept-building and to eliminate empirical perceptions from mathematical proofs. Naturally, being convincing is not a sufficient criterion: many convincing proofs have turned out to be invalid.

That a proof can be formalised means that it can be associated with already established mathematical knowledge. With respect to formalisation, Tymoczko was in harmony with formalism, which sees the formalisation of proofs to be the principal criteria for validating a proof. He was also in harmony with logicism that made explicit the rules of inference that were to be applied for building up mathematics upon a foundation of logic. However, to some extent in accordance with intuitionism, Tymoczko's point was that formalisation is not the only criteria that is operating within the mathematical practice. At times, formalisation might not even be a necessary requirement for a proof to be accepted as being valid.

That a proof should be surveyable means that it should be possible to make a step-by-step check of the proof. A mathematical proof cannot include holes and jumps. This has been taken as a given: all the proofs required for completing the logicist programme can be checked in detail; all formalist proofs can be checked; and so can all proofs in the intuitionist version of mathematics. But the computer-based proof of the four-colour theorem cannot be subjected to a step-by-step check. Maybe it can in theory, but not in any real-life mathematical practice.

From a mathematical point of view, the four-colour theorem is now considered to be solved, and computer-based proof has become common in mathematical practice. However, such practices bring about new philosophical challenges. The whole concept of what counts as an adequate mathematical proof needs to be rethought. The issues about a proof being convincing, possible to formalise, and survivable always need to be carefully addressed.

8.3 A New Philosophy

The book *New Directions in the Philosophy of Mathematics*, edited by Tymoczko (1986), includes both the article "Some Proposals for Reviving the Philosophy of Mathematics" by Hersh, and Tymoczko's own paper "The Four-Color Problem and Its Philosophical Significance". The book sets the course towards the philosophy of mathematics for which Hersh was searching.

An important manifestation of this philosophy, now known as the philosophy of mathematical practice, is found in the book *The Philosophy of Mathematical Practice*, edited by Paolo Mancosu. In the Introduction, Mancosu (2008) highlights:

> The contributions presented in this book are [...] joined by the shared belief that attention to mathematical practice is a necessary condition for a renewal of the philosophy of mathematics. We are not simply proposing new topics for investigation but are also making the claim that these topics cannot effectively be addressed without extending the range of mathematical practice one needs to look at when engaged in this kind of philosophical work. Certain philosophical problems become salient only when the appropriate area of mathematics is taken into consideration. (p. 2)

The philosophy of mathematical practice recognises that mathematical practice is eclectic and includes a variety of approaches, techniques, tools, forms of communication, conditions for publication, and opportunities for application. There are no *a priori* criteria for what to call a mathematical practice. As a consequence, the philosophy of mathematical practice answers the question "What is mathematics?" by saying that mathematics is mathematicians' practice.

This practice might include the use of diagrams in mathematical reasoning and proving. Kenneth Manders (2008) investigates diagram-based geometric practices in order to provide a route towards "rigorous diagram-based reasoning practice". According to formalism, such reasoning should be banned from mathematics. As has been argued by David Hilbert (2021) in *The Foundation of Geometry*, geometric illustrations are not essential to geometry. As had been observed with respect to Euclid's *Elements*, the use of geometric figures might bring us to assume something that needs to be properly proved. The use of diagrams invited spatial intuition to become a legitimate part of mathematical reasoning, which according to formalism is a threat to mathematical rigour. However, mathematical practice demonstrates something different: as pointed out by Manders, diagram-based reasoning constitutes an important part of contemporary mathematical practice.[4]

Since Tymoczko discussed the concept of proof with particular reference to the computer-based proof of the four-colour theorem, the use of computers has become an essential and accepted part of the mathematicians' practice. Jeremy Avigad (2008) discusses the use of computers in mathematical inquiries, which have consequently begun to blend empirical and formal processes. The idea of carrying out systematic computer-based experiments in mathematics not only adds to mathematical practice, it also establishes a new perspective on mathematical theory-building. Karl Popper (1963, 1965) has characterised the logic of scientific development as being a dialectic between conjectures and refutations. Popper principally had empirical sciences in mind. Imre Lakatos (1976) reworked Popper's general ideas with reference to mathematics, and described the logic of mathematical development as a dialectic between proofs and refutations. However, with the introduction of the computer into mathematical research practice, the difference between the logic of scientific discovery and the logic of mathematics discovery becomes hazy. In 1742 Christian Goldbach made the conjecture that now bears his name. The Goldbach conjecture states that any even number bigger that 2 can be written as the sum of two primes. We have $4 = 2 + 2$, $6 = 3 + 3$, $8 = 3 + 5$, $10 = 5 + 5$, and $12 = 5 + 7$. We can continue and get to $100 = 47 + 53$, $102 = 5 + 97$, $104 = 3 + 101$. And we can move further on: $1{,}000{,}000 = 17 + 999{,}983$, $1{,}000{,}002 = 19 + 999{,}983$, $1{,}000{,}004 = 43 + 888{,}861$. (Some even numbers can be the sum of different sets of prime numbers as for instance: $24 = 5 + 19$; and $24 = 11 + 13$.). Until now nobody has found proof of Goldbach's conjecture, nor has anybody found a counterexample. Through the use of computers, systematic attempts to find a falsification have been made, but so far this has shown that up to 4,000,000,000,000,000,000 the conjecture is valid.

Visualisation in mathematics, both with respect to proving and discovering, has now become an established and integral part of mathematical research practice.[5] Visualisations in mathematics have been condemned by formalism, with the argument that they are breaches through which uncertainties might sneak into mathematical theorising. But the use of computers has changed the situation completely—something that can be illustrated by the identification of Costa's minimal surface. A minimal surface is defined in terms of particular mathematical properties, and for many years it was assumed that only the plane, the catenoid, and the helicoid had such properties.[6] However, in 1982 Celso Costa presented a formula describing a potential fourth candidate. It was difficult to prove Costa's conjecture, as the formula was extremely complex. From 1983 to 1985, David Hoffman and William Meeks studied Costa's formula, and they applied a computer graphic to identify properties of the surface described by the formula. This identification made it possible to prove

[4]For a discussion of visualisation in mathematics, see also Giardino (2017, 2020).

[5]See Giaquinto (2008).

[6]For a characteristic of minimal surface, see Ravn and Skovsmose (2019).

Costa' conjecture. In this case, the use of the computer made it possible to create a dialectic between visualisation and proving. Such a dialectic—being different from both a dialectic between conjectures and refutations, and a dialectic between proofs and refutations—has now become an established component of mathematical research practice.[7]

In 1979, Hersh set out to provide a new beginning for the philosophy of mathematics, and the richness of the philosophy of mathematical practice demonstrates that he has been successful. The dominance of the Three Magi is broken, and the philosophy of mathematics has become closely related to mathematicians' practice.[8]

8.4 Limitations and Impacts

Mathematics was glorified by logicism, formalism, and intuitionism. Although the philosophy of mathematical practice distanced itself from those Three Magi, it similarly contributed to the glorification of mathematics.

The philosophy of mathematical practice is rooted in mathematical practice, and whatever limitations this practice might include, they are duplicated by the philosophy of mathematical practice. Like the Three Magi, the philosophy of mathematical practice treats mathematics as an isolated phenomenon that can be addressed unconnected to controversial socio-political issues. The philosophy of mathematics practice is rich with mathematical details, which makes it inaccessible for non-mathematicians. At times, it appears to be a philosophy written *by* mathematicians and *for* mathematics. The philosophy of mathematical practice addresses mathematics according to its own self-created standards. It contributes to a deceitful image of mathematics as deserving to be glorified.

Like the Three Magi, the philosophy of mathematical practice inserts mathematics into an ethical vacuum. This is illustrated by the paper "Mathematics and Ethics", where Hersh (1990) presented what he found to be the ethical issues related to mathematics. He paid particular attention to the professional conduct of mathematicians. To do so is in accordance with much discussion of professional ethics, such as that related to medicine. According to Hersh, mathematicians' conduct concerns their relationships with "staff, students, colleagues, administrators, and ourselves" (p. 21). The American Mathematical Society has provided ethical guidelines, which for instance highlights the importance of publishing without delay, correcting errors,

[7] For a discussion of experimental mathematical practices, see Bendegem (2020), Carter (2020), and Schlimm and Gonzáles (2020).

[8] Many have contributed to the philosophy of mathematical practice: Ferreirós (2016) discusses the very conception of mathematical practice; Corfield (2003) highlights the importance of turning the philosophy of mathematics "real". Hafner and Mancosu (2008) discuss interpretations of mathematical explanations; and Detlefsen (2008) addresses the notion of purity. For a broad overview see Carter (2019). See also Sriraman and Hersh (2021).

avoiding plagiarism, and giving credit to colleagues.[9] In line with such interpretations of professional conduct, Hersh highlighted that when we consider research in pure mathematics, "the ethical component is very small" (p. 22).

I look at the ethics related to mathematics in a quite different way to Hersch. To me, the discussion of professional ethics is just a small element in a much bigger panorama of ethical issues. For me, the discussion of ethics and mathematics concerns the social impact of doing mathematics; it concerns the ambiguous ways that mathematics is involved in the modern project; it concerns the wonders and horrors that can be produced through mathematics; and it concerns how socio-political priorities might form mathematical research priorities and, consequently, mathematical practices. I use the term "ethics" broadly, to refer to reflections on socio-political issues and to struggles for social and environmental justice.

The book *The Philosophy of Mathematical Practice* (Mancosu, 2008) addresses many issues, but contains nothing concerning socio-political questions related to mathematics. We can also observe that the book includes 16 chapters, all of which are written by men. Maybe the philosophy of mathematical practice provides a new beginning, but it embraces and preserves the male dominance in the philosophy of mathematics.

To our visitors from Far Away, we answered the question "What is football?" by showing them a football match between two teams from the Brazilian Serie A. But could we instead have gone to the beach in Copacabana and looked at the many football games going on there? And what about showing our visitors how children play football in the streets and alleyways?

The philosophy of mathematical practice answers the question "What is mathematics?" by referring to mathematicians' practice. But this is far from the only practice one could refer to. There are many other practices rich in mathematics—the practices of shop assistants, bank assistants, engineers, tailors, and street sellers. By defining mathematics in terms of professional mathematicians' practice, one assumes some paradigmatic preconceptions about the nature of mathematics. One assumes that behind the variety of mathematical practices there exists something that can be called "real mathematics", and that mathematicians' practice reflects this "real mathematics" better than any other mathematical practice. However, the diversity of ethnomathematical studies does not corroborate this assumption that there exists some kind of "real mathematics". One can imagine that a philosophy of music might suggest defining music as musicians' practice, but it would be a severe limitation if the musicians are always assumed to be playing in a symphonic orchestra. Like music, mathematics can be practised in a huge variety of forms and by many different groups of people.

Although it has paradigmatic limitations, the philosophy of mathematical practice is the most substantial trend in the philosophy of mathematics today. It has inspired the formulation of social constructivism as a philosophy of mathematics, as

---

[9]The European Mathematical Society has published similar guidelines. See *Ethics in Mathematics* https://en.wikipedia.org/wiki/Ethics_in_mathematics

undertaken by Paul Ernest (1998). The philosophy of mathematical practice itself is in constant development, and the 3200+ page *Handbook of the History and Philosophy of Mathematical Practice*, edited by Bharath Sriraman (2021), is a documentation of this. One of the contributions, "The Ethics of Mathematical Practice", is written by Paul Ernest (2021). To the best of my knowledge, this is the first paper that addresses ethical issues with explicit reference to the philosophy of mathematical practice. Ernest brings the discussion of ethics much further than Hersh did, and by doing so, he opens up a new and important dimension of the philosophy of mathematical practice. It is precisely this dimension that is crucial to a critical philosophy of mathematics.

The purpose of Part II of this book was to acknowledge the technical and analytical profoundness of established philosophies of mathematics, which at the same time create a socio-political isolation of mathematics. This leads to a deceitful glorification of mathematics as a unique form of knowledge, being pure and absolute.

For a critical philosophy of mathematics, it is important to recognise the position of established philosophies of mathematics, and at the same time to point out their limitations. A critical philosophy of mathematics tries to prevent any glorification of mathematics, as well as any absolutism. It sees mathematics as a human construction that is tentative, temporary, and in need of further elaboration. A critical philosophy of mathematics acknowledges that mathematics is part of many different practices, and that it is an illusion to assume the existence of any "real mathematics". It sees mathematics as a social construction integrated with other socio-political constructions, some of which might be highly controversial in nature. A critical philosophy of mathematics tries to bring mathematical practices, as well as established philosophies of mathematics, out of their self-created ethical vacuum.

8.5 Bingo in Wonderland

Even after being expelled from Cantor's Wonderland, we can still become fascinated by the landscape of huge numbers, and let me illustrate that with a short story.

Imagine that we have all died. But this is not going to be a sad story. Imagine that all of us have gone to Paradise, and that Paradise is just as we were told it was when we were children. The sky up there is light blue, the sun is shining, and each of us is sitting on a small pink cloud with a perfect fit. We listen to pleasant harp music. We all enjoy being in Paradise.

After a while we want to take a look around. We have to make eternity go past. We enter the big Bingo Hall. Up there, God is sitting, drawing numbers from a big bag. The bag contains all the natural numbers. He reads the numbers with a clear voice. We look around. Apparently nobody is reacting. Sometimes we hear a loud voice: "Shake the bag!". And God does so.

Next to us sits a person concentrating on the card in front of him. How does it all work? It is all quite simple, he answers. If God mentions a number on your card, you win right away. We go to buy a card, and we ask for a card with the numbers from 1 to 100. Gabriel, who is in charge of selling, finds the one we want. How much is it? It is 5 Euros, and we pay.

We sit down and listen, concentrating on the game, but God does not mention any of the numbers on our card. After a while, we go back to talk to Gabriel. We want to buy a different card: one with the numbers from 1 to 1,000,000. Gabriel finds us one. How much is it? It is 5 Euros.

We sit down again, listening for our number, but God does not mention any of the numbers on our new card. We go back again to Gabriel, who explains than even though there are many numbers from 1 to 1,000,000, there are so many numbers bigger than 1,000,000 that our chance of winning is practically 0%. We take a deep breath and explain that we now want to buy a card with all of the numbers from 1 up to 1,000,000! Sure, Gabriel knows that 1,000,000! means 1X 2X 3 X... X 999,999 X 1,000,000. But immediately we change our mind: we want a card with the numbers from 1 and up to 1,000,000!!!!!!!!!! We put 10 exclamation marks after each other. It is 5 Euros, Gabriel tells us.

We sit down yet again and listen out for a number on our plate to be called, but God does not mention any of our numbers. We go back to Gabriel and ask him: "Does God know the number of sand grains in the Sahara?" "Sure, he knows everything". Gabriel mentions the number. It takes some time, but time is no problem in eternity. Here, I just refer to this number as *S*. We tell Gabriel that we want a card with the numbers from 1 up to.

$$1,000,000,000,000,000!!!!!!!!!!!!!!!!!!!! \ \ldots \ !!!!!!!!!!!$$

where the number of exclamation marks is *S*. Gabriel gives us the card. It is 5 Euros.

We still do not get a number from our card called, but Gabriel gives the same explanation as he did previously: even though there are many numbers from 1 up to this new big number, there are so many numbers bigger than it that our chance of winning is practically 0%.

We return to play the game, but after a while we are yet again back to Gabriel. What was the last number that God drew from the bag? Gabriel mentions the number, which I will just call G_1. We want a card with the numbers from 1 up to G_1. Sure, we know the price, here is another 5 Euros. Gabriel nods, looking a bit puzzled.

Soon we are back to Gabriel again. "What was the number that God drew from the bag this time?" We just call it G_2, and we want a card with the numbers from 1 up to G_2. Still, we do not win, but each and every time we go back to Gabriel and ask what the last number was that God drew from the bag. In this way we get the sequence:

$$G_1, G_2, G_3, \ldots G_n, \ldots$$

They are all numbers drawn randomly from the bag containing all the natural numbers. Each time we pay the 5 Euros. We never win. It appears that a sequence

$G_1, G_2, G_3, \ldots G_n, \ldots$ of randomly selected natural numbers is ever increasing. But is that to be expected?

We leave the heavenly Bingo Hall and go back to our small pink clouds with the perfect fit. We look at each other, perhaps a bit disheartened. We have to sit here. We have to sit here for many years. Many more than 1,000,000,000,000,000!!!!!!!!!!!!!!!!!!!!! ... !!!!!!!! years. This is eternal life.

References

Appel, K., & Haken, W. (1978). The four-color problem. In L. A. Steen (Ed.), *Mathematics today: Twelve informal essays* (pp. 153–180). Springer.

Avigad, J. (2008). Computers in mathematical inquiry. In P. Mancosu (Ed.), *The philosophy of mathematical practice* (pp. 302–316). Oxford University Press.

Bendegem, J. P. (2020). Experiments in mathematics: Fact, fiction, or the future? In B. Sriraman (Ed.), *Handbook of the history and philosophy of mathematical practice*. Springer.

Carter, J. (2019). Philosophy of mathematical practice: Motivations, themes and prospects. *Philosophia Mathematica, 27*(1), 1–32.

Carter, J. (2020). Experimental mathematics in mathematical practice. In B. Sriraman (Ed.), *Handbook of the history and philosophy of mathematical practice*. Springer.

Corfield, D. (2003). *Towards a philosophy of real mathematics*. Cambridge University Press.

Detlefsen, M. (2008). Purity as an ideal of proof. In P. Mancosu (Ed.), *The philosophy of mathematical practice* (pp. 179–197). Oxford University Press.

Ernest, P. (1998). *Social constructivism as a philosophy of mathematics*. State University of New York Press.

Ernest, P. (2021). The ethics of mathematical practice. In B. Sriraman (Ed.), *Handbook of the history and philosophy of mathematical practice*. Springer.

Ferreirós, J. (2016). *Mathematical knowledge and the interplay of practices*. Princeton University Press.

Giaquinto, M. (2008). Visualizing in mathematics. In P. Mancosu (Ed.), *The philosophy of mathematical practice* (pp. 22–42). Oxford University Press.

Giardino, V. (2017). Diagrammatic reasoning in mathematics. In L. Magnani & T. Bertolotti (Eds.), *Springer handbook on model-based science* (pp. 499–522). Springer.

Giardino, V. (2020). Diagrammatic proofs in mathematics: (Almost) 20 years of research. In B. Sriraman (Ed.), *Handbook of the history and philosophy of mathematical practice*. Springer.

Hafner, J., & Mancosu, P. (2008). Beyond unification. In P. Mancosu (Ed.), *The philosophy of mathematical practice* (pp. 151–178). Oxford University Press.

Hersh, R. (1979). Some proposals for reviving the philosophy of mathematics. *Advances in Mathematics, 31*, 31–50.

Hersh, R. (1990). Mathematics and ethics. *Humanistic Mathematics Network Journal, 5*(9), 20–23.

Hilbert, D. (2021). *Foundations of geometry* (E. J. Townsend, Trans.). Altus Classics.

Körner, S. (1968). *The philosophy of mathematics*. Hutchinson University Library.

Lakatos, I. (1976). *Proofs and refutations*. Cambridge University Press.

Mancosu, P. (Ed.). (2008). *The philosophy of mathematical practice*. Oxford University Press.

Manders, K. (2008). Diagram-based geometric practice. In P. Mancosu (Ed.), *The philosophy of mathematical practice* (pp. 65–79). Oxford University Press.

McEvoy, M. (2020). Revising "The New 4CT Problem". In B. Sriraman (Ed.), *Handbook of the history and philosophy of mathematical practice*. Springer.

Popper, K. R. (1963). *Conjectures and refutations: The growth of scientific knowledge*. Routledge and Kegan Paul.

Popper, K. R. (1965). *The logic of scientific discovery*. Harper and Row.

Ravn, O., & Skovsmose, O. (2019). *Connecting humans to equations: A reinterpretation of the philosophy of mathematics*. Springer.

Schlimm, D., & Gonzáles, J. F. (2020). Mathematical experiments on paper and computer. In B. Sriraman (Ed.), *Handbook of the history and philosophy of mathematical practice*. Springer.

Shapiro, S. (2000). *Thinking about mathematics: The philosophy of mathematics*. Oxford University Press.

Shapiro, S. (Ed.). (2005). *The Oxford handbook of philosophy of mathematics and logic*. Oxford University Press.

Sriraman, B. (Ed.). (2021). *Handbook of the history and philosophy of mathematical practice*. Springer.

Sriraman, B., & Hersh, R. (2021). The history and philosophy of mathematical practice: From origin to natural historians/philosophers: A conversation. In B. Sriraman (Ed.), *Handbook of the history and philosophy of mathematical practice*. Springer.

Tymoczko, T. (1979). The four-color problem and its philosophical significance. *The Journal of Philosophy, 76*(2), 57–83.

Tymoczko, T. (Ed.). (1986). *New directions in the philosophy of mathematics*. Birkhäuser.

Part III
Indefinite Constructions

Chapter 9
Indefinite with Respect to Concepts and Proofs

Abstract A critical philosophy of mathematics considers mathematics as being indefinite. Mathematical constructions are not heading towards definite formats; they are tentative and always open to change. In this chapter I concentrate on showing mathematics as being indefinite with respect to concepts and proofs. I illustrate this with reference to the changing roles of the notion of infinitesimal. Before the nineteenth century the term infinitesimal was used liberally, but in the nineteenth century it became a concern to ensure rigour in mathematical reasoning, and to get the infinitesimals under control. The notion of function is indefinite; in fact, it has not always been part of the mathematical vocabulary. Neither Newton nor Leibniz operated with the notion of function. The very word function was first used by Bernoulli, and Euler made some further clarifications of the notion. The notion of infinity has, in its own way, brought about huge controversies in mathematics. The notion is indefinite, and this indefiniteness has a huge impact on what to consider as a valid mathematical proof.

Keywords Constructivism · Intuitionism · Indefinite construction · Indefinite with respect to concepts · Indefinite with respect to proofs · Proof · Infinitesimal · Function · Infinity · Mathematical rigour

When launching his philosophy of mathematical practice, Ruben Hersh suggested we look at mathematics as perpetually being under construction. I am going to follow up on this idea when formulating a critical philosophy of mathematics.

What does a mathematical construction site look like? Maybe one could compare it with a construction site for a building, where many workers are milling around engaged in tasks: some use a hod to bring bricks up the ladder; some prepare the mortar; some work at the top of the building; some paint the rooms; and some install a bath. However, maybe it is more appropriate to think of a mathematical construction site as being not like that of a building, but more like the construction of a whole city. Many buildings, houses, roads, bridges are under construction simultaneously. Some buildings are becoming extremely tall; some remain low level; some are located in the shadows of the tall buildings; and some look like temporary shelters. Some parts of the construction sites are abandoned, and the half-finished buildings

O. Skovsmose, *Critical Philosophy of Mathematics*, Advances in Mathematics Education, https://doi.org/10.1007/978-3-031-71375-0_9

gradually get more and more dilapidated. Some older buildings are simply pulled down in order to make space for new constructions.

My formulation of a critical philosophy of mathematics takes place in steps. In Part I, I pointed out that mathematics is not an isolated epistemic phenomenon, but is involved in all spheres of life. I also pointed out that the modern project includes profound ambiguities. Both realisations need to be recognised by a critical philosophy of mathematics. In Part II, I explored the modern philosophies of mathematics, and showed how their focus on intrinsic ontological and epistemological issues produces a false image of mathematics as being objective, neutral, and encompassing absolute truths. Part III, which I am initiating now, concentrates on showing that mathematical constructions are indefinite. In using this word, I am claiming they are not accomplished once and for all according to a pre-defined blueprint, but are always tentative and preliminary, always ready for refinement, pregnant with further possibilities. If mathematics were a pre-defined structure that could only assume one form; any critique of mathematics would be superfluous. However, when mathematics is recognised as being indefinite, critique becomes important. In Part IV, I will formulate a performative interpretation of mathematics, which establishes an intimate connection between mathematics and action, and consequently between mathematics and ethics. Any actions, including mathematics-based actions, are in need of being critically addressed. Consequently, ethics becomes an important dimension of a critical philosophy of mathematics.

I am going to argue that mathematics is indefinite with respect to concepts and proofs. Mathematical concepts are changing, and so are what is considered to be proper mathematical proofs. Mathematics is indefinite with respect to topics and applications. A mathematical theory might be developed without having any particular application in mind, but unexpected applications might subsequently arise.[1] Mathematics is indefinite with respect to power; it might come to serve all kinds of interests.[2] Mathematics is indefinite with respect to culture; it can be encultured in many different ways, which can provoke conflicts concerning which enculturation to consider the "genuine" one.[3] In the present chapter, I concentrate on exploring how mathematics is indefinite with respect to concepts and proofs.

9.1 Fallibilism

Since Plato, support for an absolutist view of mathematics has been repeatedly reiterated. Gottlob Frege embraced Platonism by seeing the world of sets as forming mathematical reality. He wanted to ensure that mathematics established absolute truths about this reality. Absolutism was advocated by formalism, but in an *if-then*

---

[1]This will be addressed more carefully in Chap. 10.

[2]This will be addressed more carefully in Chap. 11.

[3]This will be addressed more carefully in Chap. 12.

format. The claim was that *if* the axioms of a theory were true, *then* its theorems would be true as well. Intuitionism represented an anti-Platonism movement: all mathematical entities are constructed, and consequently so are mathematical truths. However, intuitionism found that if mathematical intuition was applied in an appropriate way, it would reach absolute truths.

In *Proofs and Refutations* Imre Lakatos (1976) confronted absolutism by proposing a fallibilist interpretation of mathematics. Fallibilism with respect to natural sciences has a solid base in the empiricist tradition, according to which any kind of knowledge is based on sense impressions. Empirical statements are always approximations to certain truths. An elaborated conception of fallibilism with respect to empirical sciences was formulated by Karl Popper (1963, 1965). He considered a scientific theory to be a guess, and a prevailing scientific theory is just a guess that until now has not been falsified. As falsification is always a possibility, Popper characterised the logic of scientific development as a dialectic between conjectures and refutations, which would never terminate in absolute truths.

Lakatos was inspired by Popper's interpretation of fallibilism, and he described mathematical development as a dialectic between proofs and refutations. By doing so, Lakatos explicitly proposed a constructivist view of mathematics. He showed how mathematical concepts become formulated and reformulated as integral parts of processes of proving. The dialectic between proofs and refutations has the form of dialogues taking place in the scientific community, which Lakatos in *Proofs and Refutations* elegantly condensed into a dialogue taking place in an imaginary classroom. Lakatos showed how the concept of polyhedron has developed over time, and how this conception integrates with formulations and reformulations of proofs of Euler's Polyhedron Theorem. Lakatos showed, for instance, how proof-generated concepts emerge as part of this dialectic process.

Lakatos has inspired me in suggesting a broader alternative to absolutism than fallibilism. In dictionaries, the word "fallibility" is described as a tendency to make mistakes and to be wrong. However, I do not think of tentative mathematical theorising as being "wrong"; it is just preliminary and tentative. By advocating that mathematics is indefinite, I want, like fallibilism, to confront any form of absolutism, whether it is rooted in logicism, formalism, or intuitionism. However, I want to do more than confront absolutism. I want to prepare for a critical conception of mathematics, and one important step in doing so is to recognise that mathematics is indefinite.

9.2 Infinitesimals

The notion of infinitesimal has played a huge role in mathematics. An overall challenge is set by the question: What is the result of adding up an infinity of infinitesimals?

The pre-Socratic philosopher Zeno presented this challenge in terms of an apparent paradox. If Achilles and the tortoise were running a race, and the tortoise had an advantage at the beginning, then Achilles would never be able to catch up with the tortoise. The reason was that Achilles first had to get to the point P_1, where the tortoise was when the race started. But during the time it would take Achilles to get to P_1, the tortoise would have been moving gradually forward to a point P_2. During the time it takes Achilles to get to P_2, the tortoise would again have moved slightly forward to a point P_3. And so on. By presenting this paradox Zeno showed that reasoning concerning infinitesimals needs careful attention.

Changes in the way of dealing with infinitesimals contributed to the change in mathematics that took place from the 18th to the nineteenth century. While a principal concern during the eighteenth century was to develop mathematical techniques and to achieve mathematical results, the main concern in the nineteenth century was to develop mathematics in more rigorous ways. The result-oriented mathematics from the eighteenth century manifests itself in the works by, for instance, Daniel Bernoulli (1700–1782), Leonhard Euler, (1707–1783), and Pierre-Simon Laplace (1749–1827). Results were obtained by combining careful calculations with audacious uses of intuition, not least with respect to infinitesimals. Judith Grabiner (1974) pointed out in her famous article "Is Mathematical Truth Time-Dependent?" that it might be questionable whether "Euler and his contemporaries would have been able to derive their results if they had been burdened with our standards of rigor" (p. 356).

The nineteenth century brought mathematics into a different social context. The industrial revolution was in its stage of powerful social, technical and intellectual development, and mathematics was centrally involved in this process. It became broadly recognised that mathematics had an important role to play in developing new technologies. Engineering competence became much in demand, and mathematics was an important component in any engineering education. While in the eighteenth century, mathematicians could concentrate on developing mathematics for mathematicians, the nineteenth century mathematicians also needed to teach mathematics to a broader audience. In France the École Polytechnique opened in 1794, and in 1804 its aim became to train both civil and military engineers. Soon similar schools opened all over Europe.

Mathematics became part of the engineering curriculum, which brought about demand for an educational organisation of mathematics. Augustin-Louis Cauchy (1789–1857) was a professor at the École Polytechnique, and in *Cours d'Analyse*, published in 1821, he presented a detailed organisation of the calculus.[4] Cauchy (1821) was careful in specifying notions such as limit, convergence, and continuity. In this way the external demand for teaching mathematics to a broader audience brought about an intrinsic reorganisation of mathematics itself. Grabiner (1974) saw this change as an example of a scientific revolution. When Thomas Kuhn (1962) coined the term "scientific revolution", he chiefly had in mind revolutions in the natural sciences, and his principal example was the move from a geocentric to a heliocentric worldview. Kuhn characterised a scientific revolution as a change of

[4]For an annotated translation of *Cours d'Analyse*, see Bradley and Sandifer (2009).

paradigm, which includes changes of taken-for-granted assumptions, changes of basic concepts, and changes of scientific procedures taken to be valid. Contrary to the natural sciences, it had been broadly assumed that mathematics did not undergo revolutionary changes. However, Grabiner argued that mathematics also demonstrates revolutionary changes, and she found that mathematics grows "not only by successive increments, but also by occasional revolutions" (p. 364). She added: "Mathematics is not the unique science without revolutions. Rather, mathematics is that area of human activity which has at once the least destructive and still the most fundamental revolutions" (p. 364).

As an illustration of the change that did take place from the 18th to the nineteenth century, we can try to differentiate the function $y = x^3$ following a pre-epsilon-delta approach.[5] In doing so, we need to compare the infinitesimal change of y with the corresponding infinitesimal change of x. The first step is to add an infinitesimal dx to the variable x and calculate the infinitesimal change of the variable y, namely dy. The calculation will be:

$$\begin{aligned} dy &= (x + dx)^3 - x^3 \\ &= \left(x^3 + 3x^2dx + 3xdx^2 + dx^3\right) - x^3 \\ &= 3x^2dx^2 + 3xdx^2 + dx^3 \\ &= dx\left(3x^2 + 3xdx + dx^2\right) \end{aligned}$$

In these calculations, we deal with dx as being a small number, and consequently the expression $(x + dx)^3$ can be calculated in the regular fashion.

Then we calculate the ratio between the infinitesimal changes dy and dx, and we get:

$$\begin{aligned} \frac{dy}{dx} &= \frac{dx(3x^2 + 3xdx + dx^2)}{dx} \\ &= 3x^2 + 3xdx + dx^2 \end{aligned}$$

In this calculation we continue to treat dy and dx as being small numbers, and dy/dx as being a regular fraction. However, we can look at the expression.

$$3x^2 + 3xdx + dx^2$$

in a different way. As dx is an infinitesimal, it can be considered to be 0, and this means that $3xdx$ and dx^2 can be considered to be 0. Consequently, we conclude:

$$\frac{dy}{dx} = 3x^2$$

[5] For a similar exemplification, see Ravn and Skovsmose (2019).

We have found that $y' = 3x^2$. The result is as we expected, but the proof of it is not up to the standard of rigour in demand today. It belongs to a pre-Cauchy paradigm.

As early as 1734, in *The Analyst,* George Berkeley (1685–1753) questioned the validity of operating with infinitesimals as sometimes being small numbers and at other times being equal to 0.[6] He expressed this lack of rigour by calling the applied reasoning "neither fair nor conclusive", and he continued:

> For when it is said, let the Increments vanish, i.e. let the Increments be nothing, or let there be no Increments, the former Supposition that the Increments were something, or that there were Increments, is destroyed, and yet a Consequence of that Supposition [...] is retained. (*The Analyst*, § XIII)[7]

Berkeley questioned the possibility of, at times, considering the increments dx to be small entities, and later considering them to be 0. With irony, Berkeley added that when we suppose the increments will vanish, then we must suppose that everything derived from this supposition will vanish as well. In other words, the whole argument will vanish along with the vanishing infinitesimals.

Mathematics did take heed of Berkeley's warning, although with almost 90 years' delay, and the very notion of infinitesimal was abandoned by mathematical theorising. There was a change in mathematical concept building and proving, which Grabiner characterised as a scientific revolution.

However, this was not the end of the story. With the development of non-standard analyses, rigorous use of the notion of infinitesimal became possible. Non-standard analysis was formulated by Abraham Robinson when he published the book *Non-Standard Analysis*. Robinson (1996) observed that, from an intuitive point of view, the use of infinitesimals makes sense and that the idea of infinitesimal quantities seems to appeal naturally to our intuition (see p. 2). He also observed that the use of infinitesimals was widespread during the formative stages of infinitesimal calculus. Robinson pointed out that formulating a theory of infinitesimals implies the introduction of "ideal numbers" that "might be infinitely small or infinitely large compared with the real numbers but which were *to possess the same properties as the latter* (p. 2, italics in the original). The task of non-standard analyses is to show that infinitesimals can be treated as such "ideal numbers" and that one can operate with them in a coherent way.

The way that the notion of infinitesimal has been first included, then excluded, and then included again in calculus illustrates what I mean by mathematics being indefinite with respect to both concepts and proofs. What are considered solid concepts and valid proofs during one period of time might be considered suspicious in another period of time. Mathematics development includes paradigmatic shifts, which signals that mathematic is indefinite.

---

[6] For a similar reference to *The Analyst*, see Ravn and Skovsmose (2019).

[7] See *The Analyst* at http://www.maths.tcd.ie/pub/HistMath/People/Berkeley/Analyst/Analyst.pdf

9.3 Function

The notion of function has not always been part of mathematical vocabulary. The notion has a history that stretches back in time, but nowhere near as long as, for instance, many geometric notions.[8] William L. Schaaf (1930), professor in mathematics education at Brooklyn College, City University of New York, assigned a particular value to the formation of the notion:

> [T]he keynote of Western culture is the function concept, a notion not even remotely hinted at by any earlier culture. And the function concept is anything but an extension of elaboration of previous number concepts – it is rather a complete emancipation from such notions. (p. 500)

The concept of function: a keynote in Western culture! Schaaf talked about Western culture, not only about Western mathematics. He ascribed extreme importance to the notion of function. He went almost as far as making the highly questionable presumption that the concept of function is a marker of Western supremacy with respect to the development of mathematics.[9]

In *The Geometry*, first published in Latin in 1637, René Descartes (1596–1650) introduced the idea of a coordinate system, which made it possible to use algebraic methods for addressing geometric problems (see Descartes, 1954). An algebraic expression such as $(x^2 + y^2 - 2x)^2 = 4(x^2 + y^2)$ corresponds to a graph in a cartesian coordinate system, as shown in Fig. 9.1, also referred to as the cardioid. However, Descartes did not talk explicitly about functions or graphs of functions. Instead, he talked of dealing with two variables that are connected by an algebraic expression. Having introduced the two-dimensional coordinate system, it is not a big step to introduce the three-dimensional coordinate system. In such a system, an expression connecting the variable x, y, and z would correspond to a graph in a three-dimensional coordinate system.

Isaac Newton (1642–1726/1727) and Gottfried Leibniz (1646–1716) have often been presented as inventors of the calculus. This might be true; it might also be an expression of Eurocentrism rather than of historical insight, since the Kerala mathematicians in India had developed similar ideas.[10] A principal focus of Newton's and Leibniz's investigations was geometric curves and movements along such curves. Newton operated with the notion of "fluxion", which refers to the instantaneous rate of change of a point "flowing" along a curve. Neither Newton nor Leibniz made use of the language of function.

The very word "function" was, however, used by Johann Bernoulli (1667–1748), who in 1718 stated: "One calls here Function of a variable a quantity composed in any manner from that variable quantity of numbers or constant quantities" (quoted

---

[8] To clarify the historical development of the notion of function, I have found Kleiner (1989), Lützen (1978), and Monna (1972) useful sources.

[9] I will address and strongly criticise this presumption in Chap. 12.

[10] We return to this point in Chap. 12.

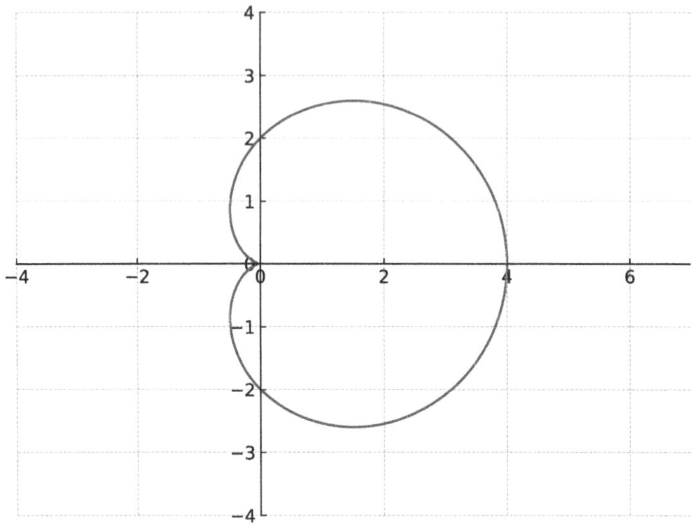

Fig. 9.1 The graph corresponding to $(x^2 + y^2 - 2x)^2 = 4(x^2 + y^2)$

by Kleiner, 1989, p. 3). This is an unclear formulation, but unquestionably the notion of function was introduced. However, it took some time before its importance became recognised. In 1748, Leonhard Euler (1707–1783) published *Introductio in Analysin Infinitorum*, where he described a function as an analytic expression composed of variables and constants. Euler followed up with examples stating that one can think of expressions such as $a + az$ and $az - 4zz$ as being functions of z. Euler also introduced the notation $f(x)$ for the value of a function f in the variable x. From Bernoulli and Euler, the concept of function became closely associated with algebraic expressions: the value of a function is determined by the values of the parameters in the algebraic expression.

In 1829 Johann Dirichlet (1805–1859) presented a function that came to be referred to as the Dirichlet function:

$$f(x) = \begin{cases} 1, & \text{when } x \text{ is rational} \\ 0, & \text{when } x \text{ is irrational} \end{cases}$$

To consider a Dirichlet function as, in fact, being a function means cutting the connection between the definition of functions and algebraic expressions. According to Dirichlet: "y is a function of a variable x, defined on the interval $a < x < b$ if to every value of the variable x in this interval there corresponds a definite value of the variable y". And Dirichlet added that "it is irrelevant in what way this correspondence is established" (Dirichlet, quoted by Kleiner, 1989, p. 10).[11] By cutting the

[11] The page number refers to the electronic version available on the net.

connection to algebraic expressions, Dirichlet expressed a radical generalisation of the concept of function.

Functions can have very different properties; one of their properties is that they can be continuous. Intuitively, a function is described as being continuous if it is possible to draw its graph without lifting the pencil from the paper. In 1861 Karl Weierstrass (1815–1897) added rigour to such an intuitive definition of continuity by systematising an epsilon-delta terminology (see Sinkevich, 2016). By means of this terminology, the definition of a function being continuous in a point x_0 can be stated in the following way:

> A function $f(x)$ is continuous in x_0 if such that for every x in the domain of f, we have.

In other words, a function f is continuous in x_0, belonging to its domain, when the value of $f(x)$ can get arbitrarily close to the value of $f(x_0)$, if x gets sufficiently close to x_0. One can then define a function as being continuous if and only if it is continuous in all points x_0 belonging to its domain. Through such an epsilon-delta definition, the intuitive understanding of continuity has been substituted with precise formulations. The specification of the concept of continuity had a tremendous impact on proofs in calculus. The whole domain of calculus was reshaped, both with respect to concept building and with respect to the rigour of proof.

For a long time, it was taken as given, also by Dirichlet, that the variable x of a function $f(x)$ must be a number. However, it became recognised that the domain of a function could be composed of many other kinds of mathematical entities. The domain of a function could, for instance, be a set of functions. Let us think of the function *dif* as meaning "to differentiate" and the domain of the function *dif* as being the set of differentiable functions. We have:

$$dif(f(x)) = f^{`}(x)$$

In case $f(x) = x^3 + x + 1$, we have.

$$dif\left(f(x) = dif\left(x^3 + x + 1\right) = 3x^2 + 1\right.$$

The function $dif(f) = f'$ has several well-known properties; for instance: $(f + g)' = f' + g'$; $(f - g)' = f' - g'$; $(f \text{ x } g)' = f'g \text{ x } fg'$; and $(f/g)' = (f'g—g'f)/g^2$.

In logic, one finds many examples of functions not dealing with numbers. In propositional calculus, the logical connectives $\neg$, $\vee$, $\wedge$, $\Rightarrow$, $\Leftrightarrow$ determine the way in which the truth value of the molecular statement is given by the truth values of its atomic parts.[12] The logical connectives can be seen as functions. The set of truth

[12] An example is shown in Fig. 5.1 in Sect. 5.3.

values V includes only two elements: $V = \{T, F\}$. The logical connective $\neg$ is a function from the set $V = \{T, F\}$ to the set $V = \{T, F\}$, defined as:

$$\neg(T) = F$$
$$\neg(F) = T$$

The connective $\vee$ is a function from V X V to V defined as.

$$\vee(T,T) = T$$
$$\vee(T,F) = T$$
$$\vee(F,T) = T$$
$$\vee(F,F) = F$$

In a similar way the connectives $\wedge$, $\Rightarrow$, $\Leftrightarrow$ can be defined as functions from V X V to V.

In the twentieth century the notion of a function as a mapping from a set to a set without restricting what sets it could be gradually became dominant. Bourbaki (2004) defined a function from a set E to a set F, in the following way: "We give the name of *function* to the operation which [...] associates with every element $x \in E$ the element $y \in F$." (p. 351, italics in the original). Bourbaki assumed a strict formalist approach to mathematics, and for him it was important to eliminate any trace of intuition from mathematical theorising, and consequently also from the definition of function. In Bourbaki's formulation there is little space left for intuition, but it is not completely squeezed out since it makes use of the expressions "operation" and "associates".

A further step in eliminating the last vestiges of intuition from the conception of function is taken by defining a function as a set of ordered pairs (x, y), where $x \in X$ and $y \in Y$, where X and Y could be any sets, and where there does not exist two ordered pairs (x_1, y_1) and (x_1, y_2) where $y_1 \neq y_2$. In this way, a function is fully specified by the set of ordered pairs of which it is composed.[13] The set F defined as:

$$F = \{(7,5), (1001,4), (\text{potato},4), (\text{moon},\text{cheese})\}$$

is a function (in so far that "potato", "moon", and "cheese" are considered well-defined objects).

The historical development of the notion of function illustrates what I mean by the notion of function being indefinite. There is no conclusive definition of the function. However, such indefiniteness might be hidden behind a curtain of didactical absolutism: many times, students of mathematics are presented with definitions

[13] Taking this step in the definition of function is in accordance with Frege's idea that the meaning of notions, relevant for logic and mathematics, can be associated to their extensions, while their intensions can be ignored. See Chap. 5.

of mathematical concepts as being a-historic and definite, including an apparently unquestionable definition of a function.

9.4 Infinity

The notion of *infinity* has caused huge controversy, and with Georg Cantor's presentation of transfinite cardinal numbers, the discussion became heated.

Cantor paid particular attention to the concept of equipotency: two sets are equipotent if and only if one can establish a 1–1 correspondence between the two sets. It is possible to establish a 1–1 correspondence between the set of natural numbers and the set of rational numbers. Cantor showed, however, that it is impossible to establish a 1–1 correspondence between the set of natural numbers and the set of real numbers. These two sets are infinite, but the cardinality of the set of real numbers is greater than the cardinality of the sets of natural numbers.

Cantor demonstrated that for any set A, finite or infinite, the cardinality of the set A is smaller than the cardinality of the powerset of A, which is the set of all subsets of A, symbolised as $\mathcal{P}(\mathrm{A})$. Thus, Cantor showed that $Card(A) < Card(\mathcal{P}(\mathrm{A}))$ for any set A. This insight brought about an infinity of sets A, $\mathcal{P}(\mathrm{A})$, $\mathcal{P}(\mathcal{P}(\mathrm{A}))$, $\mathcal{P}(\mathcal{P}(\mathcal{P}(\mathrm{A})))\ldots$ with higher and higher cardinalities: $Card(\mathrm{A}) < Card(\mathcal{P}(\mathrm{A})) < Card\,(\mathcal{P}(\mathcal{P}(\mathrm{A}))) < Card(\mathcal{P}(\mathcal{P}(\mathcal{P}(\mathrm{A})))) < \ldots$ In this way, Cantor opened up the landscape of transfinite cardinal numbers, and we have $Card(N) < Card(\mathcal{P}(N)) < Card\,(\mathcal{P}(\mathcal{P}(N))) < Card(\mathcal{P}(\mathcal{P}(\mathcal{P}(N)))) < \ldots$ Cantor showed that $Card(N) < Card(R)$, where $Card(N)$ is referred to as $\aleph_0$, and $Card(R)$ as $\aleph_1$.

One could ask whether there exists a set with a cardinality greater than the cardinality of the set of natural numbers, $\aleph_0$, and smaller than the cardinality of the set of real numbers, $\aleph_1$. Cantor assumed that no such in-between cardinality exists, which is known as Cantor's continuum hypothesis. Cantor could not prove it, and later it was shown that this was for good reason. Kurt Gödel (1940) and Paul Cohen (1963) showed that the continuum hypothesis is independent of the axiomatisation of set theory, meaning that one is free to assume that the continuum hypothesis can be added as an additional axiom to the axiomatisation of set theory, but also that its negation can be added as an additional axiom. Whatever addition one might make, the resulting system will be consistent.

As part of the intuitionist programme, restrictions were made with respect to mathematical proof. According to intuitionism, one cannot make indirect proofs, since the excluded middle is not assumed to be of general validity. One cannot prove a theorem T, by showing that assuming $\neg T$ leads to a contradiction. To show that the cardinality of the powerset $\mathcal{P}(\mathrm{A})$ of a set A is greater than the cardinality of the set A, Cantor used indirect proof. Consequently, intuitionism has discarded his whole theory of transfinite cardinal numbers.[14]

[14] The proof is outlined in Sect. 5.6.

David Hilbert confronted the intuitionist programme. He wanted to preserve the Wonderland that Cantor had created, which he also referred to as Cantor's paradise. Hilbert found it crucial to ensure that both direct and indirect proof could be used in mathematics. In 1925, Hilbert gave a lecture about infinity, which was published the next year; later it was published in English as "On the Infinite" (Hilbert, 1983). Hilbert gave the lecture after Brouwer had presented his critique of formalism, but before Gödel had presented his two incompleteness theorems. The first incompleteness theorem states that if a mathematical theory is so extensive that it includes the theory of the natural numbers, then it will be incomplete in the case it is consistent. The second incompleteness theorem states that, in order to prove the consistence of a theory, one needs to use forms of reasoning at least as complex as the forms of reasoning applied in the theory itself. However, in 1925, Hilbert set the metamathematical ambitions without clipped wings.

In "On the Infinite" Hilbert (1983) made the following programmatic statement: "The goal of my theory is to establish once and for all the certitude of mathematical methods" (p. 184). He wanted to ensure this certitude by demonstrating that mathematical theories were both complete and consistent; furthermore, he wanted to ensure this insight through metamathematical demonstrations that made use of only well-defined and uncontroversial "finite methods". In metamathematical demonstrations, Hilbert did not want to make use of indirect proof. He wanted to make sure that everybody, including the intuitionists, would feel compelled by the metamathematical results. Hilbert never did manage to specify what "finite methods" might include, and in the end Gödel demonstrated that the aims of metamathematics were illusionary.

One of Hilbert's (1983) concerns was to ensure that the notion of infinity could be used in mathematics. He states that "the definitive clarification of *the nature of infinity* [. . .] is needed for *the dignity of the human intellect* itself" (p. 185, italics in original). Hilbert is not only talking about mathematics, but about human intellect in general. This is an extremely strong formulation with respect to the importance of a mathematical concept. Hilbert saw the notion of infinity as a cornerstone in human knowledge building. It was needed for "the dignity of the human intellect".

Even though Hilbert put together a dedicated defence of Cantor's Wonderland, and simultaneously attacked intuitionism, the intuitionist position with respect to infinity has found much support. In the philosophy of mathematics, this support has taken the form of finitism according to which it does not make sense to operate with mathematical entities composed of infinitely many elements. The consequence is, for instance, that one cannot talk about natural numbers as being a set $N = \{1, 2, 3, \ldots\}$ and count this set as being a mathematics object. According to finitism, one can talk about the sequence of natural numbers, and about the possibility of always being able to add one more number, $n + 1$, to a sequence of natural numbers: $1, 2, 3, \ldots, n$. In this sense, one can talk about the potential infinity of natural numbers, but not about any actual infinity of natural numbers.[15]

[15] For a broader discussion of finitism, see Tiles (2004) and Ye (2011).

The notion of infinity is indefinite. It can always undergo reconstruction. It is indefinite in the same way as the notions of infinitesimal and function are indefinite.

Through all historical periods, human beings have made remarkable architectural constructions: the Mayan pyramids, the Egyptian pyramids, the Chinese Wall, the Parthenon, the Pantheon, St Peter's Basilica, the Taj Mahal Mausoleum, and the Kremlin. They are solid constructions made of bricks and mortar. Mathematics is also a human construction, where the building blocks are concepts and the mortar is the proofs. Traditionally, the mathematical architecture has been thought of as being entirely solid: what has been proved remains proved for ever. However, I have tried to show that the image of mathematics as being solid is deceptive. The material used for mathematical constructions—the concepts and the proofs—is not long-lasting. The apparent solid mathematical architecture is made of perishable material. In this sense, mathematics is indefinite.

References

Bourbaki, N. (2004). *Elements of mathematics: Theory of sets*. Springer.

Bradley, R. E., & Sandifer, E. C. (2009). *Cauchy's Cours d'analyse: An annotated translation*. Springer.

Cauchy, A.-L. (1821). *Cours d'Analyse*. De L'Imprimerie Royale.

Cohen, P. J. (1963). The independence of the continuum hypothesis. *Proceedings of the National Academy of Sciences of the United States of America., 50*(6), 1143–1148.

Descartes, R. (1954). *The geometry of René Descartes*. (David Eugene Smith & Marcia L. Latham, Trans.). Dover Publications.

Euler, L. (1748). *Introductio in analysin infinitorum*. Free Download, Borrow, and Streaming: Internet Archive.

Gödel, K. (1940). *The consistency of the axiom of choice and of the generalized Continuum hypothesis with the axioms of set theory*. Princeton University Press.

Grabiner, J. V. (1974). Is mathematical truth time-dependent? *The American Mathematical Monthly, 81*(4), 354–365.

Hilbert, D. (1983). On the infinite. In P. Benacerraf & H. Putnam (Eds.), *Philosophy of mathematics: Selected readings* (2nd ed., pp. 183–201). Cambridge University Press.

Kleiner, I. (1989). Evolution of the function concept: A brief survey. *The College Mathematics Journal, 20*(4), 282–300.

Kuhn, T. (1962). *The structure of scientific revolutions*. Chicago University Press.

Lakatos, I. (1976). *Proofs and refutations*. Cambridge University Press.

Lützen, J. (1978). Funktionsbegrebets udvikling fra Euler til Dirichlet. *Nordisk Matematisk Tidsskrift, 25/26*(1), 5–32.

Monna, A. F. (1972). The concept of function in the 19th and 20 centuries, in particular with regard to the discussions between Baire, Borel and Lebesgue. *Archive for History of Exact Sciences, 9*, 57–84.

Popper, K. R. (1963). *Conjectures and refutations: The growth of scientific knowledge*. Routledge and Kegan Paul.

Popper, K. R. (1965). *The logic of scientific discovery*. Harper and Row.

Ravn, O., & Skovsmose, O. (2019). *Connecting humans to equations: A reinterpretation of the philosophy of mathematics*. Springer.

Robinson, A. (1996). *Non-standard analysis*. (Revised ed.). Princeton University Press.

Schaaf, W. L. (1930). Mathematics and world history. *The Mathematics Teacher, 23*(8), 496–503.
Sinkevich, G. I. (2016). On the history of epsilontics. *Antiquitates Mathematicae, 10*, 183–204.
Tiles, M. (2004). The philosophy of set theory: An historical introduction to Cantor's paradise. *Dover*. Publications.
Ye, F. (2011). *Strict finitism and the logic of mathematical applications*. Springer.

Chapter 10
Indefinite with Respect to Topics and Applications

Abstract Mathematics is indefinite with respect to topics. This is illustrated by the Modern Mathematics Movement, which brought about revolutionary changes to the mathematics curriculum. These changes were facilitated by an incoherent combination of political preoccupations, military interests, economic priorities, and intrinsic mathematical ideas. That mathematics is indefinite with respect to applications is illustrated by the radical change in research related to algorithms. The *Entscheidungsproblem* is a specific metamathematical problem, whose solution presupposed that the notion of algorithm was clearly specified. This specification was done by means of the Turing machine, a conception tremendously important in the development of computing. That mathematics is indefinite with respect to applications is also illustrated by the development of modern cryptography, which refers to cryptography based on computing. In the development of modern cryptography, number theory plays a huge role. Results from apparently harmless pure number theory have acquired massive economic and also military interests.

Keywords Mathematics as indefinite · Modern mathematics movement · Royamount seminar · Algorithm · *Entscheidungsproblem* · Turing machine · Cryptography · Modern cryptography · Encryption · Decrypting · Enigma · Plaintext · Cyphertext · Number theory

At first sight, the mathematics curriculum appears fixed, laid down by authorities beyond the classroom. Neither students nor teachers can change it. The curriculum appears definite, but this is an illusion. The curriculum, in terms of what learners are presented with, not only differs from country to country, but also changes even more drastically over time.

Thomas Kuhn (1962) has shown how scientific revolutions are part of the development of natural sciences. Such revolutions might include changes in the overall worldview—for example, in what to consider a reliable scientific method, and what to consider an exemplary study. When looking at the development of the curriculum in mathematics, one can also witness evolutions taking place over time with respect to what are considered crucial issues to address and techniques to

O. Skovsmose, *Critical Philosophy of Mathematics*, Advances in Mathematics Education, https://doi.org/10.1007/978-3-031-71375-0_10

master. This illustrates what I mean by mathematics being indefinite with respect to topics.

When I was at school, geometric constructions played an important role. We were meticulously introduced to the following procedure: first, we had to draw a small figure showing the information given in advance. Second, we had to construct the figure using a ruler and a compass. Third, we had to write a short text presenting, step by step, what we had done. If we look at geometry from a Platonic perspective, such a procedure makes sense. Precisely by making a ruler-and-compass construction, one demonstrates that the geometric figure exists in the world of ideas.

In Denmark, as well as in many other countries, such constructions dropped out of the mathematics curriculum long ago. We are faced with a paradigmatic shift in the mathematics curriculum, including a change in the overall conception of geometry and of what we are to consider an exemplary geometric task. This shift was initiated by the Modern Mathematics Movement and illustrates what I mean by mathematics being indefinite with respect to topics.

Mathematics is also indefinite with respect to applications. Here, I refer to the fact that new and unexpected applications of a mathematical theory might appear. Gottfried Leibniz (1703) showed the possibility of using a binary number system when doing calculations. At the time, this observation was treated as a mathematical curiosity, without any relevant applications. However, with the invention of the electronic computer, Leibniz's observation achieved tremendous relevance beyond mathematics itself.

10.1 The Modern Mathematics Movement

A revolutionary change to the teaching of mathematics took place from the eighteenth to the nineteenth century, when mathematics became a teaching discipline for both civil and military engineering.[1] This revolutionary change was propelled forward by a combination of political interest and military priorities. A similar combination of factors lies behind the revolutionary changes in the mathematics curriculum initiated in the 1950s, and developed during the 1960s and 1970s as the Modern Mathematics Movement.

The Second World War had come to an end, and what Winston Churchill in 1946 baptised as an iron curtain had divided Europe. The Cold War created a new reality which was dominated by the political and military tensions between East and West. The military power of the Soviet Union had been clearly recognised by the West; this power was essential for defeating Nazi Germany. It was also assumed that this military power was principally due to the overwhelming size of the Soviet army, and not to particular technological advancements. In 1957, however, the Soviet Union launched the first orbiting satellite, Sputnik, which came as a shock to the West. The

[1] See Sect. 9.2.

launching of Sputnik demonstrated that the Soviet Union was well in front of the West in developing technology with military significance.

The Sputnik shock brought mathematics education into sharp focus. It became recognised that in order to catch up with the Soviet Union, technological development in the West needed a boost. In order for this to happen, it was broadly assumed that mathematics education, as the foundation for science and technology, needed to be improved. These political and military rationales were accompanied by an economic rationale. An acceleration of technological development was not only essential for developing military power, it was also essential for economic development.

The seminar *New Thinking in School Mathematics* took place in 1959 at Royaumont, an old abbey north of Paris. Here, a group of mathematicians met to discuss possible improvements in the teaching of mathematics. This seminar was organised and financed by the Organisation for European Economic Co-Operation (OEEC).[2] It was a powerful political and economic organisation that had an interest in developing mathematics education, considered crucial for ensuring technological innovations, economic growth, as well as military power.

The Royaumont Seminar was introduced with a lecture by the mathematician Marshall H. Stone, who highlighted the connection between the teaching of mathematics and the development of technology. He stated that the "teaching of mathematics is coming to be more and more clearly recognized as the true foundation of the technological society which it is the destiny of our time to create" (OEEC, 1961, p. 18). One could have imagined that Stone's lecture would have set the agenda for the seminar. He had expressed a clear connection between the overall political aims associated with the development of technology and the teaching of mathematics by calling the teaching of mathematics the "true foundation of the technological society". One could have imagined that the seminar would concentrate on clarifying what mathematical topics might contribute to the further development of technology, and how to make these topics part of an overall mathematics curriculum. One might have imagined that the whole area of applied mathematics received full attention. But this was not what happened.

Right after the opening lecture, the seminar took an abrupt turn; the political, economic, and technological agenda was promptly ignored. The subsequent lecture "New Thinking in School Mathematics" was given by the French mathematician Jean Dieudonné, and his lecture became the principal reference for the whole seminar. The title of his lecture was also chosen as the title of the report from the Royaumont Seminar (OEEC, 1961). Dieudonné had been one of the founding

[2]The founding members of the OEEC were Austria, Belgium, Denmark, France, Greece, Iceland, Ireland, Italy, Luxembourg, Netherlands, Norway, Portugal, Sweden, Switzerland, Turkey, and the United Kingdom. Later OEEC was transformed into the Organisation for Economic Co-Operation and Development (OECD).

members of the Bourbaki group, and he strongly advocated the structuralist view of mathematics, as cultivated by Bourbaki.[3]

Dieudonné (1961) started his lecture by stating that he wanted to examine the secondary school mathematics curriculum from the perspective of the university. He wanted to point out what mathematical background it would be desirable for students to have when entering university studies, what mathematical background the students actually have, and how to improve the existing situation. Dieudonné was deeply concerned about the huge gap between what was desirable, and what was actually the case.

According to Dieudonné, reforms in secondary school mathematics had to take place in order to prepare students adequately to take university-level mathematics. His principal stance was that the actual content of university mathematics must determine the curriculum for school mathematics in such a way that the gap between desirable and actual could be reduced. Dieudonné pointed out that the curriculum of the secondary schools had to be reorganised "in order to eliminate any undue waste of time and to absorb as much as possible of the burden now resting entirely on the university" (OEEC, 1961, p. 34).

The "burden" resting on the universities is the requirement for them to fill in the gap between the mathematical knowledge that the students actually have when they enter the universities, and the mathematical knowledge that they could have had if the school curriculum been organised in a more appropriate way. Dieudonné had a clear idea about the nature of this gap:

> In the last 50 years, mathematicians have been led to introduce not only new concepts but a new language, a language which grew empirically from the needs of mathematical research and whose ability to express mathematical statements concisely and precisely has repeatedly been tested and has won universal approval. But until now the introduction of this new terminology has (at least in France) been steadfastly resisted by the secondary schools, which desperately cling to an obsolete and inadequate language. And so when a student enters the university, he will most probably never had heard such common mathematical words as, set, mapping, group, vector space, etc. (p. 34)

Dieudonné referred to the development of mathematics that had taken place over the previous 50 years. During this period, mathematics had changed radically; new concepts had come to play a significant role for mathematical theory building. Dieudonné simply found school mathematics to be outdated.

Dieudonné condensed the reform that he had in mind with the slogan: *Euclid must go!* For centuries, Euclid's *Elements* had been taken as the paradigmatic way of organising and presenting mathematics. In particular, it had been assumed that geometric constructions had to play a crucial role. However, according to Dieudonné, the *Elements* no longer served as an exemplary way of organising mathematical knowledge. The *Elements* belonged to a paradigm of the past, using obsolete and inadequate language that secondary schools "desperately cling to". It

[3] See Bourbaki (2004a, b, c) and Dieudonné (1970). For a discussion of structuralism and Bourbaki, see Sect. 6.4.

did not belong to the paradigm that had been established within the previous 50 years. The *Elements* belonged in a mathematical museum, and so did the exercises I was so carefully instructed to do when in school. Euclid must go in order to make space for new concepts and a whole new mathematical language. The slogan "Euclid must go!" encapsulated the structuralist view on mathematics. Instead of cultivating historical traditions, school mathematics had to prepare students to move around in the actual architecture of mathematics.

As a result of the Royaumont Seminar, a group of 16 mathematicians were brought together to elaborate guidelines for a new curriculum in secondary school mathematics. The result of their work was the *Synopsis of Modern Secondary School Mathematics* (OECD, 1961). Informed by this synopsis, new textbooks were produced in many countries, first for secondary school mathematics, but soon for school mathematics in general. The Modern Mathematics Movement became a reality.

The Modern Mathematics Movements encompassed apparently contradictory components. It was initiated by the critical situation of the West being behind the Soviet Union in the development of military technology. Huge political and military interests were committed to solving this crisis. The technological capacity of the West had to be expanded, which in turn was associated with huge economic interests. However, it was not applied mathematicians who took over the scene with explicit concerns for augmenting a technological capacity, but instead pure mathematicians assuming a structuralist approach to mathematics. The Modern Mathematics Movement assigned a new powerful role to pure mathematicians: they were not only responsible for developing mathematics through their research, but also for defining the new school mathematics curriculum.

It was soon recognised that the recommended "modern" curriculum did not work in educational settings. The Modern Mathematics Movement turned into a broadly recognised educational fiasco.

10.2 Algorithms

That mathematics is indefinite with respect to applications I will illustrate with reference to a particular metamathematical problem. Initial attempts to clarify this problem brought about a new insight that had an impact which reached far beyond the scope of metamathematics.

The metamathematical programme, as originally formulated by David Hilbert, was ambitious. The aspiration was to address mathematics in a systematic way and to show that, in mathematics, paradoxes could not emerge. In order to demonstrate this, mathematics should be formalised, and the subsequent formalised theories should be investigated in a rigorous way. Crucially, it was necessary to clarify whether or not a formalised theory was consistent and complete. A theory is *complete* if for any formula F, belonging to the theory, it can be shown that either

F or $\neg F$ is a theorem. It is *consistent* if for any formula F, belonging to the theory, it can be show that F and $\neg F$ cannot both be theorems.[4]

When a theory has been formalised, it seems straightforward to ask whether a certain formula can be demonstrated as being a theorem or not. Given a particular formalised theory, one can presume that there is an algorithmic procedure for clarifying this question. The identification of such an algorithm has been referred to as the *Entscheidungsproblem,* or the decision problem. One could presume that the *Entscheidungsproblem* could be solved for all formalised mathematical theories. One can, in fact, presume that there exists a universal algorithm which can be applied to any kind of formalised theories, to show whether a given formula is a theorem or not.[5]

We know that the *Entscheidungsproblem* can be solved for propositional calculus. It can be shown that a formula is a theorem in propositional calculus if and only if it is a tautology. As it is an algorithmic procedure to produce a truth table for a propositional formula and to check if it is a tautology or not, the *Entscheidungsproblem* is solved.

Yet can the *Entscheidungsproblem* be solved for all formal theories? In addressing this question, it becomes important to establish a general conception of what to consider as an algorithmic procedure. Mathematics is shot through with numerous examples of algorithms. With respect to the quadratic equation $ax^2 + bx + c = 0$, it is possible to decide in an algorithmic way the number of solutions the equation $ax^2 + bx + c = 0$ has: none, one, or two, and to calculate the solutions. But in what way can one formulate a general definition of an algorithm?

Alan Turing (1912–1954) addressed the *Entscheidungsproblem* in its general form, and he created a clear conception of what to consider as an algorithmic procedure by describing an imaginary machine that he called an "automatic machine" or, for short, an "a-machine" (see Turing, 1937). Before long the a-machine became known as the Turing machine.

The Turing machine can be visualised as a tape reader with an infinitely long tape passing through the reader. The tape is divided into cells c_0, c_1, c_2. . ., and each cell contains one and only one symbol from a specific list of symbols: s_0, s_1, . . . s_n. One of the symbols represents the blank cell. This list constitutes the alphabet of the Turing machine. One and only one cell can be in the reader at any one time; the reader can register the symbol written in the cell; and it can erase the symbol and substitute it with another symbol from the list. At any one time, the Turing machine can be in one of a finite set of possible stages: q_0, q_1, . . . q_n. One of these stages is the passive one, and when the machine gets to this, it stops, having completed what it

---

[4]Gödel (1962) demonstrated through his incompleteness theorem that the metamathematical aspirations could not be fulfilled. See Sect. 6.2.

[5]In the tenth of Hilbert's unsolved mathematical problems, which he presented in his 1900 lecture at the International Congress of Mathematicians in Paris, he formulated a decision problem with respect to the solution of Diophantine equations. For an English translation of Hilbert's lecture, see Reid (1995). In 1926, Hilbert and Ackerman presented the *Entscheidungsproblem* in a more general way (see Hilbert and Ackerman, 1950).

was asked to do. The other stages are the active ones, and when a machine is in an active stage, it will read the corresponding symbol on the tape. Based on what symbol it reads and what stage it is in, the machine will write a new symbol from the alphabet in the cell, move the tape beneath the reader, and enter a new stage.[6]

The Turing machine seems to be quite a simple mechanism, but it is all-encompassing. Whatever kind of mathematical algorithm one can imagine, it has an equivalent in a Turing machine. Consequently, a Turing machine provides a general characteristic of an algorithmic procedure.

By means of the Turing machine, Turing addressed the *Entscheidungsproblem.* If the algorithm was the one that determines whether a particular formula is a theorem or not in a given formal system, then solving the *Entscheidungsproblem* for that formal system would be equivalent to showing that the associated Turing machine would come to a stop. To solve the general *Entscheidungsproblem* would mean showing that there is a universal Turing machine that determines whether Turing machines in general will come to a stop. Turing's trick was to apply such an imagined universal Turing machine to itself. This brought about a contradiction, and by applying an indirect proof, Turing concluded that no such universal Turing machine could be constructed. As a consequence, he concluded that there is no general solution to *Entscheidungsproblem.*

Turing did solve the metamathematical *Entscheidungsproblem*, although he did so by showing that the problem did not have a general solution. In doing so he created the notion of the Turing machine, which turned out to have an extremely explorative power. Since any mathematical algorithm can be represented by a Turing machine, the machine provides an abstract model of an electronic computer, which basically speaking is a conglomerate of interacting mathematical algorithms. A computer exists in a multitude of versions, and new versions are constantly being created. However, a computer can still be considered equivalent to a Turing machine. Or rather: a powerful computer can be considered a good approximation of a Turing machine. The concept of the Turing machine has become part of the theoretical foundations of computer science, crucial for clarifying what can be achieved by computational means.

The creation and application of the Turing machine is an example of applications of mathematics being indefinite. The *Entscheidungsproblem* was formulated with respect to a particular metamathematical issue. It was unpredictable that such an

---

[6]More formally, to define a Turing machine one needs to define the set of stages $Q = \{q_0, q_1, \ldots q_n\}$; the set of symbols $S = \{s_0, s_1, \ldots s_n\}$; the set of cells $C = \{c_0, c_1, \ldots c_n \ldots\}$, which can be infinite; the initial stage of the machine, q_{initial} to be chosen among the elements of Q; the terminal stage, q_{term} (the passive stage) also to be chosen among the element of Q; and a transition function T that prescribes what is happening when the Turing machine is in stage q_x, the cell c_y is in the reader, and the machine reads the symbol s_z. Thus $T(q_x, c_y, s_z) = (q_{\text{new-x}}, c_{\text{new-y}}, s_{\text{new-z}})$, where $q_{\text{new-x}}$ refers to the new stage of the machine, $c_{\text{new-y}}$ to the new cell in the reader, and s_{new} to the new symbol written in the cell c_y. For a more accurate formal definition of the Turing machine, see, for instance, Chandran (2012) and Sermutlu (2020).

approach intended to address the *Entscheidungsproblem* would have a substantial impact on the further development of computer science.

10.3 Cryptography

Number theory has often been considered as a mathematical discipline without any practical applications: the epitome of pure mathematics. Modern cryptography refers to a cryptography based on computing, and for developing modern cryptography, number theory has come to play a huge role. This also illustrates what I mean by mathematics being indefinite with respect to applications.

As an illustration of pre-modern cryptography, one can think of a cryptographic system based on a specific edition of a particular book. If we want to code a text, we can refer to a letter by a triple of numbers (x, y, z), where x refers to a page of the book, y to a line on the page, and z to the number of a letter in the line. In this way an encrypted text comes to appear as a sequence of ordered triples. Apparently, this system is difficult to compromise, but it has at least two serious weaknesses: the processes of both encrypting and decrypting are laborious, and the code is immediately broken, if the book that is used is identified.

During the Second World War, the Germans were using an advanced mechanism for encrypting and decrypting messages between operational command and the troops, referred to as the Enigma machine. The encoding and decoding processes were able to be rapid because the Enigma machine, invented during the 1920s by the German engineer Arthur Scherbius, was designed as a typewriter. When typing a message, the machine did the encryption automatically through a system of interchangeable gears, and the encrypted text could then be transmitted. The receiver using another Enigma machine could type in the coded text, and then the original text would appear. A resetting of the gears in the Enigma machine was regularly undertaken, which made it even more difficult to break the code.

But the code was broken. In Bletchley Park, Buckinghamshire, a group of experts was brought together, and under the direction of Alan Turing the Enigma messages were decrypted, eventually quickly and efficiently.[7] This had a huge impact on the outcome of the Second World War, and because the Germans shared their Enigma machine with the Japanese, the breaking of the Enigma code also supported the Allies' victory in the Pacific.[8] The Enigma machine was an advanced cryptographic system based on mechanical procedures. It also turned out to denote the end of the mechanical approach to cryptography.

[7] In 1934 the Enigma code had already been broken by a group from the Polish Cipher Bureau; see Rakus-Anderson (2003).

[8] It has been claimed that Turing's work shortened the war by 2 years and saved 14 million lives, leading the screenwriter of the film *A Beautiful Mind,* Akiva Goldman, to quote the mathematician John Nash: "Mathematicians won the war."

In order to illustrate how number theory is deeply involved in modern cryptography, I will briefly outline some of the ideas of the so-called *public key system*. The possibility of establishing such a system was presented in the article "New Directions in Cryptography" by Diffie and Hellman (1976). Since this initial beginning, the use of mathematics in cryptography has grown explosively; however, concentrating on this initial example of modern cryptography is sufficient to illustrate my point that new and unexpected applications of mathematics can occur.

Let me introduce some terminology used in cryptography.[9] The text that is going to be coded is called the "plaintext", referred to as P. The process of encrypting the plain text is referred to as E. The cryptic text called the "ciphertext" is $C = E(P)$. The process of decoding a ciphertext we refer to as decryption, D. The system of cryptography should naturally be designed in such a way that:

$$D(C) = D(E(P)) = P$$

In modern cryptography the processes of encryption and decryption are based on mathematical algorithms and computing. This means that a huge message can be encrypted and decrypted with speedy efficiency. An apparent consequence is that breaking the code can be done with the same speedy efficiency. The principal question becomes how we can design an algorithmic way of encoding and decoding, which cannot easily be compromised.

In the public key system, the approach for solving this problem hinges on the application of a "trap-door function". This is a function f, where it is easy to calculate $y = f(x)$ for a given x, but difficult to calculate x given y and f. Number theory provides a simple example of a trap-door function. If the function f is defined as $f(x, y) = xy$, and f is acting on huge prime numbers, then even if we know the result $f(x, y)$ of the multiplication, it is extremely difficult to identify which two prime numbers x and y have been multiplied.[10] In order to see that the function f is a trap-door function, it is important to know the extreme difficulties associated with factorising the product of two very large prime numbers. Why is it so difficult? Can we be sure that it is always difficult for all huge primes x and y? Could one imagine a discovery of a new algorithm that makes it relatively easy to do the factorisation? The discovery of such an algorithm could be extremely important for breaking the code in the public-key system. Once the factorisation is completed, the code is broken.

Let me indicate why this is so. We let EK_1 refer to encryption using a key we call K_1, and DK_2 refer to decryption using a key called K_2. In this context the notion

[9] In the following, I draw on Skovsmose and Yasukawa (2009). For an introduction to modern cryptography, see Schröder (2009).

[10] Schröder (2009) states: "At present, a 200-digit number that is the product of two 100-digit numbers cannot be factorised in any reasonable time … In fact, not so long ago, the most efficient factoring algorithms on a very fast computer were estimated to take 40 trillion years, or 2000 times the present age of the universe" (p. 131).

"key" refers to the device that is used for completing the encryption and the decryption. With this symbolism we have:

$$C = EK_1(P)$$
$$P = DK_2(C)$$

We can think of the plaintext P as being a number.[11]

One can imagine a system where the key for encryption is the same as that for decryption: $K_1 = K_2$. Such a system would presuppose that the sender and the receiver of the cryptic message both know the key, and keep it secret. If the key was generated by the sender, he or she must communicate it to the receiver. If the key gets captured by a third part, then the whole process is compromised.

The public key system operates with two different keys. One key, K_1, is the public one, which is known by the sender and the receiver, and which could be known by the third party without compromising the system. K_2 is the private key, and it is only known by the receiver; not even the sender needs to know the private key since it is only used for decrypting the ciphertext. In the public key system, there is no need to communicate the private key or to share it with anybody, which makes the system more secure. Naturally, it is important that it is impossible to identify the private key K_2, even when knowing the shape of the public key K_1 and the ciphertext C.

To construct such a public-key system, one needs to create a procedure for generating the public key K_1, and the private key K_2. We assume that we have identified two large prime numbers p and q, and consequently we can calculate their product $n = pq$. By drawing on a range of number theoretical results, such as Euler's Theorem,[12] one can generate two other numbers, e and d, which can be used for defining the public and the private key in the following way:

$$K_1 = (e, n)$$
$$K_2 = (d, n)$$

meaning that the two keys are ordered pair of numbers.

With this definition of the keys, the process of encrypting a plaintext P and decrypting the corresponding ciphertext C can be defined as:

$$C = P^e (\text{mod } n)$$
$$P = C^d (\text{mod } n)$$

[11] The reason is that a text can be coded as a number using, for instance, the approach applied by Gödel (1962).

[12] This theorem states that for all numbers a and n, which are relatively prime to each other and for which $n > 0$ and $0 < a < n$, it is true that $a^{\phi(n)} \equiv 1 \ (\text{mod } n)$. The function $\phi(n)$ is known as Euler's totient function.

In other words, encrypting a plaintext P is done by raising P to the power of e, and dividing the result by n. The rest will then be the ciphertext C. Decrypting a ciphertext C is done by raising C to the power of d, and dividing the result by n. The rest will then be the plaintext P.

As parts of the public key $K_1 = (e, n)$, both e and n are well known. Consequently, a plaintext can be encoded without including any secret steps. Part of the private key $K_2 = (d, n)$, namely n is known, but d is not. This means that if d is identified, then the code is broken. As d has been constructed, one will identify d immediately if one can factorise n and thereby reveal the two prime numbers p and q that have been multiplied as $n = pq$.

This initial beginning of modern cryptography illustrates what I mean by mathematics being indefinite with respect to applications. Mathematics can find applications that have not previously been recognised. Much research in mathematics is directed towards specific applications, while other research is considered "pure". However, there is no predictability with respect to potential applications of mathematics, pure or not. Number theory—an apparently pleasant and peaceful research domain—has come to contribute in an essential way to the launching and steering of rockets with precision, and the making of a heavy rain of bombs. Being "pure" is a momentary characteristic of any mathematical theory.

References

Bourbaki, N. (2004a). *Elements of mathematics: Elementary theory*. Springer.

Bourbaki, N. (2004b). *Elements of mathematics: Theory of sets*. Springer.

Bourbaki, N. (2004c). *Elements of mathematics: Functions of real variable*. Springer.

Chandran, R. (2012). *Turing machine*. Remya Chandran.

Dieudonné, J. (1961). New thinking is school mathematics. In OEEC (Ed.), *New thinking in school mathematics* (pp. 31–49). OEEC.

Dieudonné, J. (1970). The work of Nicolas Bourbaki. *The American Mathematical Monthly, 77*, 134–145.

Diffie, W., & Hellman, M. (1976). New directions in cryptography. *IEEE Transaction on Information Theory, IT-22*(6), 641–654.

Gödel, K. (1962). *On formally undecidable propositions of principia Mathematica and related systems*. Basic Book.

Hilbert, D., & Ackerman, W. (1950). *Principles of mathematical logic*. Chelsea Publishing Company.

Kuhn, T. (1962). *The structure of scientific revolutions*. Chicago University Press.

Leibniz, G. (1703). *Explanation of binary arithmetic*. http://www.leibniz-translations.com/binary.htm.

OECD. (1961). *Synopsis of modern secondary school mathematics*. OECD.

OEEC. (1961). *New thinking in school mathematics*. OEEC.

Rakus-Anderson, E. (2003). The brains behind the Enigma code breaking before the World War II. In B. Booss-Bavnbek & J. Høyrup (Eds.), *Mathematics and war* (pp. 83–102). Birkhäuser.

Reid, C. (1995). *Hilbert*. Springer.

Schröder, M. R. (2009). *Number theory in science and communication: With applications in cryptography, physics, digital information, computing, and self-similarity* (5th ed.). Springer.

Sermutlu, E. (2020). *Automata, formal language, and Turing machines*. Emre Sermutlu.

Skovsmose, O., & Yasukawa, K. (2009). Formatting power of 'mathematics in a package': A challenge for social theorising? In P. Ernest, B. Greer, & B. Sriraman (Eds.), *Critical issues in mathematics education* (pp. 255–281). Information Age Publishing.

Turing, A. M. (1937). On computable numbers, with an application to the Entscheidungsproblem. *Proceedings of the London Mathematical Society, s2-42*(1), 230–265.

Chapter 11
Indefinite with Respect to Power

Abstract The close connection between mathematics and power is expressed in many ways. One way is the formation of the digital panopticon that is making use of big data. We are all potentially open to being observed as the data about our behaviour is collected, stored, and made available for analysis. Big-data analysis demands new mathematical and statistical techniques; it is of huge interest to financial markets, for social control, and for the military. Economic decision-making is an expression of the mathematics-power amalgamation. Algorithmic procedures can ascribe a credit score to a person, and based on this score a bank can judge whether the customer should be offered a loan or not. The FICO system was an original way to do so, but recently e-score systems have been created that make use of big data. Social power concerns the governing of and conditions for democratic life. The crucial question is whether further mathematisation of society tends to support or obstruct democratic life. Mathematics contributes to research on enhancement of military efficiency, and much of this research is directly financed by the military.

Keywords Power · Panopticon · Digital panopticon · Big data · FICO · Credit score · e-score · Democracy · Mathematisation of society · Echo chamber · Military power · Cryptography · Dispersion theory

Knowledge and power get integrated when some knowledge becomes institutionalised as being "genuine". By claiming that what is written in the Bible must be taken as absolute truth, religious dogmatism becomes an example of such an amalgamation. The Catholic Inquisition castigated any form of blasphemy, and it did so with a brutality that included torture and execution.

The scientific revolution made space for genuine knowledge—or so it was assumed by Modernity. Superstition and dogmatic ideas were supposedly eliminated, and scientific knowledge was glorified as the epitome of genuine knowledge. However, one critique of the modern conception of knowledge is that it is illusory to assume knowledge to be time-independent. It has been pointed out that the modern image of knowledge is an ideological stereotype. Knowledge that has been considered objective, neutral, and time-independent has been revealed to be precast

O. Skovsmose, *Critical Philosophy of Mathematics*, Advances in Mathematics Education, https://doi.org/10.1007/978-3-031-71375-0_11

according to preconceptions that could well be false. Knowledge, including scientific knowledge, can be seen as a "regime of truth", and like other regimes it may be challenged, and maybe overthrown.[1] What might be nominated as scientific knowledge during a certain period is a manifestation of dominant power structures.

The intimate connection between knowledge and power also concerns mathematics. Applications of mathematics can be extremely powerful, and can be of direct political, economic, bureaucratic, and military interests. As mathematics is indefinite with respect to topics and applications, it is also indefinite with respect to power. Mathematics can associate itself with any form of power structure; it is not particularly selective. Mathematics seems to be ready to serve whoever and whatever.

11.1 Digital Panopticon

In a series of letters written in 1786 and 1787, Jeremy Bentham delineated the possibility of the construction of a prison with a panopticon format that would make it possible for a small group of guards to oversee and control a larger group of prisoners.[2] He spent much time refining the plan of his prison, as he hoped that the British government would eventually adopt it. This never happened and the plan never became a reality; the idea of the panopticon was largely ignored.

In 1949 when George Orwell published *Nineteen Eighty-Four*, the idea of the panopticon received a new form of expression. In the dystopia that Orwell (1987) described, people are controlled by Big Brother by being watched. People cannot escape being watched, as every household has to install a TV set, which can be used in two ways: people can watch TV programmes, but people can also be watched from behind the screen. Big Brother can see what is taking place in each and every household, but Big Brother himself can never be seen. The TV sets can never be turned off, so being watched by Big Brother is a permanent and threatening possibility. Big Brother might not be watching, but one never knows. By means of the Big Brother metaphor, Orwell expressed a principal feature of power: namely that the executer of power remains invisible, and those subjected to power know that they might be observed, but not when nor by whom.

In 1975, Michel Foucault published the original French version of *Discipline and Punish: The Birth of the Prison* (Foucault, 1979), where he gave Bentham's notion of panopticon a new significance. A panopticon does not only refer to the possible architecture of a prison; it refers to a general pattern of social control; it refers to prisons, schools, and social institutions in general. While the Bentham panopticon is of an architectonical format, and the Orwell panopticon is of a mechanical format, the Foucault panopticon is of an organisational format.

[1] Foucault (2000) uses the expression "regime of truth".

[2] See Bentham (2011).

Due to computer technology, it has become possible to establish a new kind of panopticon that I will refer to as the *digital panopticon*. Such a panopticon is installed in many workplaces today. It can register automatically when an employee enters the workplace and opens the computer. Whatever he or she is doing can be registered, day after day after day. When and where the employee moves around the internet, it can be registered. Maybe nobody will ever see the registered information, but it is collected and placed in a data bank. The data might be scanned by algorithmic devices that identify cases that have been stipulated as not-normal or non-productive. In the digital panopticon everyone is potentially being observed.

When the data collection is enormous, we are now talking about *big data*. For dealing with such data, conventional mathematical and statistical methods are not adequate. Big data demands new ways of processing and analysing, and the mathematical and statistical techniques used to do so are in exponential development. Big-data analysis is of huge interest for research; for financial institutions; for decision-making with respect to investments and economic risk-taking; for the design of marketing programs; for implementing healthcare programmes; and for designing military strategies. Big data has become merchandise. The collection and analysis of big data is also important for many environmental programmes. For instance, images from satellites have been combined with other kinds of information to provide clearer images of the deforestation taking place in the Amazon rainforest.

The concept of Big Brother has undergone metamorphosis. Orwell coined the term as an ironic warning against the formation of totalitarianism by means of extensive and intensive observation. The concept connected total control with permanent observation. Big Brother as a TV show was created by the Dutch media magnate John de Mol Jr. Big Brother was shown for the first time in 1999, and immediately it was copied in various countries around the world. Through this TV show, we can all watch a group of nice-looking young people: how they perform in carefully designed competitions, how they live together; how they talk to each other; how they make intrigues; how they smile; how they get upset. We can see where they sleep; who they sleep next to; how they might embrace each other. We can see everything that the producers of the programme use to stimulate the viewers' curiosity without provoking strong negative reactions.

We, the viewers, become Big Brothers. We might be dealing with a relatively innocent entertainment programme, but the popularity of the Big Brother programme also symbolises a change in attitudes towards being observed: it becomes trivialised and normalised.

In the Introduction to this book, I mentioned that I was listening to music when writing, and I am also doing that now, as I draft this paragraph. I am using the application Spotify. When Miriam and I celebrated New Year's Eve and watched the TV programme showing 2023 in retrospect, Miriam showed me that Spotify had also made a retrospect of the music that I had been listening to during 2023. Spotify had, for instance, provided a ranking of the musicians that I had been listening to the most. The retrospect showed how many minutes I had been listening to music during the whole year, and also which month I had spent most time listening. It had collected much more information on my musical and listening habits than I was aware of.

I was surprised that this kind of information about me had been collected. I had not asked for this information to be collected; it had been collected automatically and secretly, meaning that mathematical algorithms had been carefully designed and implemented for doing the job. For what purpose? An enormous amount of similar data must have been collected about millions and millions of people—in 2024 it is estimated that Spotify had 574 million users. For what purpose is it collected? We can only speculate. Such big data might be solely collected for our personal amusement at each new year. However, Spotify might have other reasons behind trawling through our preferences. It might be used to develop commercial strategies. It might be used to characterise different profiles of music consumers. The information might be generated to be sold to interested companies and institutions. The digital panopticon seems to be without limitations.

I could have become upset and angry when I found out that my music priorities had been so carefully monitored and registered, but I just shrugged my shoulders. It was, somehow, only what I should have expected. I also became a bit curious to see the ranking of my favourite choices. However, shrugging the shoulders might be one more example of docile passivity. We might be dealing here with yet another example of the production of big data that could be used for any purpose, now or in the future. We might be dealing with one more addition to the totalitarian panopticon—made possible due to hidden and unobserved mathematical techniques—that is potentially a peril to democratic life.

11.2 Economic Power

People need to buy houses, cars, and fridges—and many need to borrow money in order to do so. They could try to get a loan from a bank, but the bank might not readily offer them a loan.

A bank does not "give" loans. It is in the business of "selling loans", which are its principal source of income. The more loans it can sell, the better. However, a client might not be able to pay back the loan, and the bank will eventually lose its money. One bank strategy could be to provide loans only to consumers with an obvious reliable source of income. However, this will very much reduce the number of potential customers. The bank has a direct economic interest in providing as many loans as possible, and simultaneously keeping the associated risk to a well-specified minimum. The aim of a bank is not to eliminate all risks entirely. To earn money, a bank needs to engage in risk-taking, but it is crucial for the bank to have precise estimations of the risks that are being taken. The bank needs to evaluate how much money might be lost and what the probability for this is. This is vital for specifying the interest rate.

A bank could evaluate the credibility of a consumer by interviewing the person and considering his or her financial situation, as shown from the consumer's bank account and from similar available sources. However, such a collection of information is time- and resource-consuming. It would be much more efficient to have available algorithmic and automatic procedures to ascribe a *credit score* to the client. Based on this score the bank could then judge whether the client should be offered a loan or not. A credit-score system would be much welcomed by the banking industry. It would make it possible to administrate a huge number of loan-offers with a degree of automaticity.

The FICO system is an answer to this need. FICO stands for Fair Isaac Corporation, which is the name of a company, founded in 1956, selling credit-score systems. The FICO credit-score system applies a mathematical model designed by the engineer William Robert Fair and the mathematician Earl Isaac, and it has been broadly applied by banks and financial institutions. It is in ongoing development; I have seen references to FICO-9. One can also find a version on the internet where one can estimate one's own credit score.

Very simplified, a credit-score system is a mathematical model M, connecting a complexity of variables. The variables can be listed as $x_1, x_2, \ldots x_n$, and the output of the model can be referred to as $M(x_1, x_2, \ldots x_n)$. The credit scores, CS, will then be $CS = M(x_1, x_2, \ldots x_n)$. When constructing a credit-score model, one needs to consider what features of reality to incorporate in the model, and what features to ignore. This is essential for defining the list of variables $x_1, x_2, \ldots x_n$. Some variables are included in the model, while others possible variables are ignored. What about the customer's age? What about gender? What about ethnicity? What about the neighbourhood where the customer is living? What about family relationships? There are very many variables one can imagine a bank would find relevant for defining a potential loan-taker's credit score. For an outsider it might be impossible to come to know what variables a credit-score system is, in fact, operating with. The selection of variables might remain confidential to the bank.

The variables in a mathematical model need to be connected, and such connections can be established through equations. In this way, the whole algorithmic machinery gets to operate. However, the way in which the variables become connected is not any predetermined thing. The connections to be made are not direct consequences of empirical observations; they are established due to simplifications, assumptions, intuitions, presumptions, and decisions. It has to be decided upon which variable should play the dominant role in a credit-score model, which should play only minor roles, and which should play none. A variable x_i is dominant in the case where a change of the value of x_i $(1 \leq i \leq n)$ gives a significant change of $M(x_1, x_2, \ldots x_n)$, while a variable x_j is less significant, if a change of its value only brings about a minor change of $M(x_1, x_2, \ldots x_n)$.

For applying the credit-score model, values have to be ascribed to the variables $x_1, x_2, \ldots x_n$. The values might be inserted directly into the model by the bank assistant, who asks the client for the relevant information. We can call such data *perceptible data*. However, today many credit-score models operate with a much less transparent way of ascribing values to variables. The value of some variables might be specified by resources which neither the bank assistant nor the client are aware of. Some data might be inserted into the model by tapping into some big-data funds. This kind of data I refer to as *concealed data.*

Instead of talking about credit scores, one now often talks about *e-scores*, referring to a measure of a person's economic credibility. The development of models for calculating e-scores has been explosive, and models have been developed drawing on big data, meaning that e-scores are partly based on concealed data extracted from digital panopticons.

In 2016 Cathy O'Neil published *Weapons of Math Destruction* which has the subtitle *How Big Data Increase Inequality and Threatens Democracy.* O'Neil (2016) is deeply concerned about the social impact of big data, and she also makes particular observations with respect to the mathematical modelling of e-scores. She writes:

> Since Fair and Isaac's pioneering days, the use of scoring has of course proliferated wildly. Today we're added up in every conceivable way as statisticians and mathematicians patch together a mishmash of data, from our zip codes and Internet surfing patterns to our recent purchases. Many of their pseudoscientific models attempt to predict our creditworthiness, giving each of us so-called e-scores. These numbers, which we rarely see, open doors for some of us, while slamming them in the face of others. (p. 143).

The systems for calculating e-scores draw on an extensive amount of concealed data; these might include the patterns of the person's surfing on the internet, their Facebook profile, and their purchases. One could imagine that even a person's music preferences could, in some way, be correlated to the person's financial credibility. An e-score system might be designed in such a way that a preference for music by black youth might reduce the person's financial credibility (and Spotify might be ready to provide the data). We might be dealing with an example of a mishmash of data included in an e-score model.

The formation of the Fourth World is an ironic feature of globalisation.[3] This formation is an expression of the social and economic exclusion that accompanies processes of globalisation. This exclusion can also be executed in terms of e-scores. Just imagine an e-score calculation operating with concealed parameters with respect to patterns of surfing on the internet and to previous purchases. What about people who have no computer and no mobile, and who do not do any internet shopping? What e-score would they get? The e-score systems might be constructed in a way that eliminates huge groups of people from economic facilities. E-score systems might contribute to the exclusion by fencing off the hyperghettoes also referred to as the Fourth World—not by barbed wire, but by numerical labelling.

The power that can be acted out through mathematical models drawing on big data might be enormous. Power might be increased, while transparency becomes reduced. We are dealing here with a direct threat to the democratic way of life.

---

[3]See Sect. 4.3 for a discussion of the notion of the Fourth World. See also Castells (1998).

11.3 Political Power

Conditions for establishing a democratic way of life are partly shaped and reshaped by mathematics brought into action, such as through the handling of big data and the massive development of artificial intelligence. I am going to refer to these phenomena as *mathematisations of society*, and the question is: Does a mathematisation of society strengthen democratic endeavours, or does it rather favour totalitarian political trends?

In addressing this question, it is important to clarify the very notion of democracy. Over time, the concept of democracy has obtained many and also apparently contradictory interpretations. Here I will draw on my own interpretation, which I try to clarify by commenting on *voting*, *fairness*, *equity*, and *deliberation*.[4]

Voting. In a direct democracy, everybody is involved in the decision-making taking place in the public space, which I also refer to as the *agora*. However, in a larger community it is impossible for everybody to get together, and consequently voting becomes a crucial act. Voting is a procedure that allows somebody else to speak in one's own name. It is a procedure for establishing a congregation whose responsibility is to ensure that decisions that are taken are expressions of the shared will of the whole population.

Could one imagine that new voting procedures might become possible due to the mathematisation of society? In the mid-1980s, I gave a course in the philosophy of technology to students in mathematics and computer science at Aalborg University. We were deeply interested in the possible social impact of the emerging computer technologies, and looking back at what we discussed at that time, I can summarise it by saying that we got it all wrong. We discussed the new possibilities for democratic life. Would revival of direct democracy become possible, even in huge communities? We could imagine that any political question could be raised in the online virtual agora, where it could be voted upon and where the results could be known instantly. In a democracy it is important that the government listens to public opinion, and computer technology made such active listening feasible. This was how we, in the mid-1980s, imagined the future.

Although the computerisation of society might bring about new ways for articulating public opinion, it also establishes new ways of implementing grand-scale behavioural modifications (see Zubofff, 2018). The formation and handling of big data establishes various forms of behaviour shaping—for example among voters. Even if the mathematisation of society makes active listening to public opinions possible, it also makes active manipulation of such opinions possible.

Fairness. Many times, voting has been considered a defining feature of a democracy, but democracy concerns much more than the existence of voting procedures. It also concerns the distribution of welfare; thus it is difficult to imagine that the

---

[4]Previously, I have discussed the concept of democracy (see Skovsmose, 1994). I have also elaborated on the notion in Part 3, "Democracy as a Challenge", in Skovsmose (2014). A recent presentation of my conception of democracy can be found in Skovsmose (2023).

existence of an extreme difference between poor and rich people is an expression of a *shared* will. The neo-liberal understanding of democracy does not consider the distribution of welfare as a condition for democratic life, but is fully focused on the freedom to speak and the right to vote. In my understanding, democracy also refers to material life-conditions, and consequently I see fairness with respect to welfare as being an important feature of democratic life.

Computers are expensive, and so are cell phones. "Getting access" has become an expensive commodity, not available to everybody. In order to get access to and operate in the labour market today, one needs to be able to "connect". If not, one might be left behind in the ever-growing pool of "disposable people". The unfair distribution of opportunities to "get access" is a threat to the democratic way of life, formed by the mathematisation of society.

Equity. In a democracy everyone should be treated equally according to the law. People cannot be treated differently with respect to ethnicity, religion, poverty, richness, sexuality, caste, or any other such categories. Equity not only concerns human rights; it also concerns opportunities in life. It concerns access to welfare and education, and it concerns access to information and knowledge. Modern information technologies make it easier for some to access information and knowledge, but more difficult for others to do so. The political economy of "getting access" concerns equity. It concerns a threat to the democratic ideals of equity brought about by the mathematisation of society.

Deliberation refers to processes unfolding in the public space; it refers to life at the agora. A public debate might include demagogic proclamations and fake news, but the existence of such phenomena is not an argument for barring groups from accessing the public debate. Whatever is taking place at the *agora*, everybody should have equal access to this. An important feature of a democracy is the argumentation that lies behind public decision-making. However, the extensive use of mathematical modelling for political decision-making removes the premises for this decision-making from the public sphere and locates them within narrow circles of technical expertise. Instead of being the result of a deliberation, political decisions become consequences of technical optimisations. Naturally the mathematical expertise could question outcomes of mathematical investigations in a broader public debate, but this is far from common. The self-imposed isolation of the mathematical expertise is highly problematic for the life at the democratic *agora*.

Kathleen Hall Jamieson and Joseph N. Cappella (2008) provide a careful study of the phenomenon of echo chambers, which refer to the way messages are magnified and propagated through the media. They see an echo chamber as a space that ensures loops with positive feedback for those who join the chamber. People that enter an echo chamber might be predisposed to certain political opinions, which get exposed and augmented. The echo in an echo chamber can be a flashy political voice, as alternative voices are excluded and explicitly discredited. An echo chamber can have a decisive influence on political opinion-making and can be pivotal for the outcome of an election. Echo chambers might well have existed in all historic periods, but the mathematisation of society now makes it possible to construct echo chambers of an enormous size and simultaneously to establish fast-working feedback loops. Echo chambers are mathematics-created technical possibilities that impede democratic deliberations.

Throughout history, totalitarianism has found many expressions, not least during the period of Modernity: Nazi Germany, Fascist Italy, and Stalinist Soviet Union being just three examples. These cases of totalitarianism are based largely on the physical domination of people. Mark Coeckelbergh (2022) offers a warning about the possible formation of an AI totalitarianism. This totalitarianism is also based on power, but of a different kind, namely the informational domination of people. Extensive registration of a population contributes to the domination of that population. This is a knowledge-based domination rooted in the digital panopticon. I am concerned about the force of the totalitarian propensities that are incised in the mathematisation of society. These propensities threaten the democratic ideals with respect to voting, fairness, equity, and deliberation.

11.4 Military Power

During his military campaigns from 1797 to 1801 in Syria and Egypt, Napoleon recognised the role that mathematics could play in multiplying military efficiency. In Paris, the École Polytechnique was opened in 1794, and once back from Egypt, Napoleon turned it into an institution for educating both civil and military engineers. Ever since Napoleon's militarisation of the École Polytechnique, the intimate relationship between mathematics and military power has developed further and further.

The First World War marked the close connection between mathematics, science, technology, industrialisation, and military power. While previously, military killing had been a kind of handcraft, now killing became industrialised. The capacity of the artillery was enhanced tremendously; it was no longer necessary to see the enemy in order to kill them. Ballistics as a mathematical discipline achieved a new significance, making killing at a distance more efficient.

From the start of the First World War in 1914 to the close of the Second World War in 1945, military engineering developed enormously, and mathematical war research expanded massively.[5] Research institutions with a focus on the military use of mathematics were opened by the military powers. Ballistics drew on, for instance, the solution of differential and integral equations, statistical prediction, and numerical methods. Aeroplane design and aeronautics drew on geometry, calculus, complex functions, and gas dynamics (which in turn applied nonlinear hyperbolic equations). Weather forecasting, applying a range of mathematical methods, achieved a new importance as military operations needed to be planned also with

[5] The connections between mathematics and the military have been discussed in *Mathematics and War*, edited by Booss-Bavnbek and Jens Høyrup (2003). Most of the following references are to articles from this book. For a panoramic view on military work in mathematics from 1914 to 1945, see Siemund-Schultze (2003).

weather conditions in mind. Cryptography was of huge military interest and drew on logic, number theory, coding theory, and statistics. The design of rockets applied a range of mathematical disciplines, including numerical analysis. The making of the atomic bomb was the result of a variety of mathematical disciplines.[6]

Andrey Kolmogorov is famous for presenting an axiomatisation of probability theory in 1933. He was also a unique applied mathematician, and during the Second World War he contributed to ballistics by developing a firing dispersion theory. Such a theory specifies the area which a shell fired from a cannon in a certain position would hit. Most likely, two shells fired from the same cannon would not hit exactly the same spot but have a certain likelihood of exploding within a certain area, depending on the inclination of the cannon, the shell, and the distance between the cannon and the target. In order for an artillery to "cover" a certain area, it is important to determine the area covered by each of the cannons in the artillery. An artillery might be positioned a long distance from the front line, while the enemy's positions might be located quite close to its own front lines, and in order to minimise this risk of killing own soldiers, it is important to have available a carefully elaborated dispersion theory. Kolmogorov's formulation of such a theory was demanded by the military of the Soviet Union soon after the country was invaded by Germany in 1941. One month later Kolmogorov presented his theory. Simultaneously he turned a firing dispersion theory part of the curriculum for studies in probability theory.[7]

In June 1943 the commander Kazuo Kamaga, from the General Staff of the Japanese army, demanded the collaboration of the Mathematics Department of the Imperial University of Tokyo in studying military cryptography. The ambition was, naturally enough, to create an unbreakable code, but soon the research concentrated on decoding procedures that would make it possible to read the enemy's messages. Cryptographic research in Japan found it important to study the mathematical work of, for instance, Leonhard Euler and Évariste Galois. Mathematical disciplines such as number theory and algebra achieved a completely new significance by contributing to this research.[8] In Germany cryptographic research had resulted in the manufacture of the Enigma machine for encrypting and decrypting. For the Allies, cryptographic research had resulted in breaking the code.

[6]In 1917 in Germany, the Aerodynamische Versuchsanstalt in Göttingen opened, financed by the military. In 1928 the Institut für Praktische Mathematik in Darmstandt opened, which contributed to the rocket project in Peenemünde. In 1915, the National Advisory Committee on Aeronautics was founded in the USA. In 1925 a mathematical division was established in the Bell Telephone Laboratories (Bell Labs), which engaged in research of direct relevance for the military. In 1938 the Ballistic Research Laboratory in Aberdeen, Maryland was established; it served as a centre for developing war technologies. In 1943 the Los Alamos National Laboratory was established as part of the Manhattan Project; it was a centre for nuclear research and was responsible for the construction of the atomic bomb. Institutes aimed at carrying out mathematical war research also opened in the United Kingdom, France, Italy, and Japan, financed completely or partly by the military.

[7]For a presentation of Kolomogorov's contribution to mathematical war research, see Shiryaev (2003).

[8]For a discussion of cryptographic research in Japan, see Fukutomi (2003).

New mathematical disciplines and research topics emerged in the aftermath of the Second World War. Mathematics had contributed to war research, but when the war was over, the new topics were not removed from the mathematical research agenda. Instead, several of the new topics became integrated into the regular university study programmes. Topics from mathematical war research were normalised. The outbreak of the Cold War and the Sputnik shock might also have contributed to this "normalising". The growth and establishment in the post-war period of such new disciplines at American universities was generously supported by military financing. In 1948 the RAND Corporation was founded with the explicit aim of providing research for the armed forces and for the defence industry. In 1946, the Office of Naval Research was established, with the aim of developing research that could enhance the power of the navy.[9]

I have illuminated the close relationship between mathematics and the military by referring to studies of the past. It would be important to refer to studies of the present, but I have not come across such studies. This is not surprising as military issues concern national security and are guarded behind closed doors. We know about the fierce treatment of whistle-blowers and the treatment of Julian Assange. The extension of mathematical war research has been broadly ignored by mathematics research and educational communities as well as by philosophies of mathematics. In order to give the world a wake-up call, we are in need of not only whistle-blowers, but of trumpetists.

Mathematics and war are closely integrated, but what about mathematics and peace? Ubiratan D'Ambrosio (2024) has carefully considered what a nonkilling mathematics might mean. As an initial step he invites us to reflect on the fact that mathematics has been an instrumental companion to historical events that we all condemn. However, he does not assume the existence of inherent malignant features of mathematics, but finds that mathematics could come to play "an important role in achieving the high humanitarian ideals of a new civilization with equity, justice and dignity for the entire human species, without distinction of race, gender, beliefs and creeds, nationalities and cultures" (no page numbering in original). In using mathematics to move towards such a utopia, one must master mathematics, but this is not enough. One also needs to establish a critical view of the potentialities and of the risks involved in using mathematics. D'Ambrosio's considerations about a nonkilling mathematics are a direct expression of the claim that mathematics is indefinite with respect to power. Mathematics might serve deplorable interests, but it need not necessarily do so. It could be a means of moving towards a utopia—a world in peace.

Architectural constructions can serve many kinds of military, political, or religious interests: Hadrian's Wall, the Great Wall of China, the Colosseum, the Egyptian pyramids, etc. The power of modern progress was celebrated by the erection of the Eifel Tower. The Olympic stadium in Berlin was built for the

[9] For a discussion of the establishment of new mathematical disciplines in the wake of the Second World War, see Kjeldsen (2003).

Olympics taking place in 1936. It was designed with a gigantic Olympic Gate, which could serve as a scene for *der Führer* to appear on. The Nazis also planned the construction of *der Volkshalle* (People's Hall) in Berlin which could also demonstrate the power of the Nazis. It was supposed to be covered by the biggest cupola ever constructed. The barbed-wire fencing of the Jewish ghettoes in German cities was a demonstration of Nazi power, and so was the construction of extermination camps for carrying out the "final solution". Another horrendous contemporary architectural construction is the Israeli wall between Israel and Palestine, located deep within Palestinian territory. Like architectural constructions, so can mathematical constructions come to serve many kinds of power. Military priorities guide investment into mathematical research, even research which, up close, might appear neutral and pure. Like architecture, mathematics is indefinite with respect to power.

References

Bentham, J. (2011). *The panopticon writings*. Edited by M. Božovič. Verso Books.
Booss-Bavnbek, B., & Høyrup, J. (Eds.). (2003). *Mathematics and war*. Birkhäuser.
Castells, M. (1998). *The information age: Economy, society and culture. Volume III, End of millennium*. Blackwell Publishers.
Coeckelbergh, M. (2022). *The political philosophy of AI*. Polity Press.
D'Ambrosio, U. (2024). Nonkilling mathematics? The ethics of mathematics in the final analysis. *Philosophy of Mathematics Education Journal*, (41).
Foucault, M. (1979). *Discipline and punish: The birth of the prison*. Penguin Books.
Foucault, M. (2000). *Power*. The New Press.
Fukutomi, S. (2003). Mathematics and war in Japan. In B. Booss-Bavnbek & J. Høyrup (Eds.), *Mathematics and war* (pp. 153–159). Birkhäuser.
Jamieson, K. H., & Cappella, J. N. (2008). *Echo chamber*. Oxford University Press.
Kjeldsen, T. H. (2003). New mathematical disciplines and research in the wake of the World War II. In B. Booss-Bavnbek & J. Høyrup (Eds.), *Mathematics and war* (pp. 126–152). Birkhäuser.
O'Neil, C. (2016). *Weapons of math destruction: How big data increase inequality and threaten democracy*. Broadway Books.
Orwell, G. (1987). *Nineteen Eighty-Four*. Penguin Books.
Shiryaev, A. N. (2003). On the defence work of A. N. Kolmogorov during World War II. In B. Booss-Bavnbek & J. Høyrup (Eds.), *Mathematics and war* (pp. 103–107). Birkhäuser.
Sigmund-Schultze, R. (2003). Military work in mathematics 1914–1945: An attempt at an international perspective. In B. Booss-Bavnbek & J. Høyrup (Eds.), *Mathematics and war* (pp. 23–82). Birkhäuser.
Skovsmose, O. (1994). *Towards a philosophy of critical mathematics education*. Kluwer Academic Publishers.
Skovsmose, O. (2014). *Critique as uncertainty*. Information Age Publishing.
Skovsmose, O. (2023). *Critical mathematics education*. Springer.
Zubofff, S. (2018). *Surveillance capitalism: The fight for human future at the new frontier of power*. Profile Books.

Chapter 12
Indefinite with Respect to Culture

Abstract The colonisation of mathematics is engraved in the version of the history of mathematics that presents it as a Western achievement. As part of the process of colonisation, Eurocentrism developed into an all-dominating perspective on the world, and the white-supremacy ideology became part of the modern outlook. The European colonisation of mathematics was accomplished by cutting away the non-Greek roots of mathematics, and by describing Greek culture as European. The Eurocentric presentation of the history of mathematics is, however, a myth that has transformed into an unquestionable given, repeated in many mathematics textbooks, including brief sketches of the history of mathematics. Ethnomathematical studies have contributed profoundly to the recognition of the multiplicities of the enculturation of mathematics. For a critical philosophy of mathematics, it is crucial to acknowledge this multiplicity and to recognise mathematics as being indefinite with respect to culture.

Keywords Colonisation of mathematics · Eurocentrism · Rafael's school of Athens · Enculturation · Racism · Scientific racism · Enculturation of mathematics · Genocide · North American genocide · Decolonisation of mathematics · Indefinite with respect to culture

It has been claimed that mathematical truths are independent of time and context, and that mathematics exists in some sublime format. In line with such absolutism, some cultures become considered carriers of genuine mathematical ideas and, in this way, to represent epistemic supremacy.

I find such claims to be without foundation; they are merely efforts to *colonise mathematics*. The colonisation of mathematics, comparable to any other form of colonisation, is engraved in presentations of the history of mathematics that portray Western cultures as the carrier of genuine mathematics. Contrary to any such hegemonic ideas, I find mathematics to be indefinite with respect to culture.

O. Skovsmose, *Critical Philosophy of Mathematics*, Advances in Mathematics Education, https://doi.org/10.1007/978-3-031-71375-0_12

12.1 Rafael's School of Athens

On many occasions I have enjoyed being a tourist in Rome. I have visited the Vatican Museum several times and each and every time I have been impressed by the power of Michelangelo's (1475–1564) paintings in the *Capello Sixtina*, showing for instance a naked Jesus demonstrating a magnificent set of muscles. I have visited *Stanze di Raffaello*, with the frescoes painted by Rafael (1483–1520). The *Capello Sixtina* and *Stanze di Raffaello* are sublime examples of high-renaissance art.

One of Rafael's frescoes is *Scuola di Atene* (School of Athens), which was painted between 1509 and 1511. The fresco depicts philosophers and mathematicians including Plato, Aristotle, Socrates, Diogenes, Pythagoras, Heraclitus, Archimedes, and Euclid. Everybody seems preoccupied: discussing, writing, copying, listening, watching. Nobody is looking in the direction of the viewer, except for a woman dressed in white, looking directly towards us with a Mona-Lisa-like smile, and Rafael himself who has sneaked into the picture.

Rafael made many personal references in the fresco. Leonardo da Vinci (1452–1519) assumes the role of Plato in the tableau. The role of Aristotle is played by Giuliano da Sangallo (1445–1516), who was a sculptor and architect who inspired both da Vinci and Rafael. Socrates appears to the left of Plato and Aristotle, and is involved in discussions with some attentive young people. Diogenes is resting in the middle of the stairs with his only two belongings: a cloak and a wooden bowl. At the bottom of the stairs Michelangelo assumes the role of Heraclitus absorbed in speculation. It was easy for Rafael to create a good resemblance of Michelangelo, who at the time was working right next door in the *Capello Sixtina*. The important role of Euclid, drawing a circle with a compass, is played by Donato Bramante (1444–1514), the architect who in 1503 was hired by Pope Julius II to plan the reconstruction of St. Peter's Basilica. Some have suggested that Bramante represents not Euclid but Archimedes. I would, however, find it surprising if Rafael had not given Euclid a prominent position in the fresco. One sees Pythagoras concentrated on making some notes. Bending forward, trying to catch what Pythagoras is writing, one sees Ibn Rušd (1126–1198), often called Averroes, who has written about a broad range of topics, including philosophy and mathematics.

Rafael's fresco is built up with a clear linear perspective, with the vanishing point placed between Plato and Aristotle. The use of a linear perspective in paintings distinguishes Renaissance art from Medieval art. Medieval art was a central component in the glorification of God, and any tableau was assumed to be seen from a God-like perspective. Emphasis on the human perspective was not a constituent part of Medieval art, but as part of the Renaissance and the accompanying humanisation, human beings assumed a new status. One expression of this is through seeing the world with the human eye, and in the most direct way this leads to the application of the linear perspective.[1]

Besides the linear perspective, Rafael's School of Athens also demonstrates a Eurocentric perspective. Rafael's School of Athens is a condensed artistic expression of seeing both philosophy and mathematics as a European achievement. It illustrates what I have in mind when talking about colonisation of mathematics.

12.2 Colonisation of Mathematics

As part of the process of colonisation, Eurocentrism developed into the all-dominating perspective on seeing the world. Extraordinarily, sea voyages resulted in the "discovery" of "new worlds", and colonisation and slave-trading became an accepted part of the world order. The very words "discovery" and "new world" are direct expressions of a Eurocentric perspective. The brutality of colonisation called for legitimisation, and in order to justify the trade in humans, racism was made an integral part of the Eurocentric worldview. European culture was claimed to be superior due to the superiority of white people when compared to "black savages". This supremacy ideology soon became an accepted component of the modern outlook.

With this worldview, what should we think of mathematics? It was well known that mathematics had been developed to extremely sophisticated levels in Mesopotamia and in Egypt. The pyramids were the most visible expression of Egyptian mathematical capabilities, while remarkable astronomic predictions testified to the Mesopotamian dominance of mathematics. How could it be that cultures relegated to being inferior, according to the European supremacy-ideology, could contribute to the development of mathematics, which was considered a sublime epistemological achievement? Provoked by this question, the supremacy ideology came to include a careful reworking of the history of mathematics, which took the form of the colonisation of mathematics.

This colonisation included two strategic manoeuvres. The *first manoeuvre* was to cut off the variety of non-European roots of mathematics. What was considered to be non-European mathematics was downgraded to being only preliminary expressions of mathematical thinking and to being empirically based rather than deductively organised. Egyptian mathematics might well have been brought into action for constructing the pyramids; specific cases of the Pythagorean Theorem could have been applied for the construction of right angles, but the general Pythagorean Theorem was not part of Egyptian mathematics. At least, so it was claimed. In a similar way, Mesopotamian mathematics was downgraded to being merely a

[1] Around 1425, Brunelleschi (1404–1472) carefully investigated the geometry of perspectives, and his architecture and studies of perspective influenced many renaissance architects and artists, including Giuliano da Sangallo, da Vinci, and Rafael.

conglomerate of empirically based laws. Other formulations of mathematics, such as Indian, Chinese, and Japanese mathematics, were largely ignored.[2]

The second manoeuvre concerned Greece. Instead of seeing Greek culture as an integral part of Egyptian and Eastern cultures, it was recast as being a European culture. The Greek language was investigated carefully, and during the 1800s German philologists engaged in comprehensive comparisons between the Greek and the German language.[3] The ambition was to document the existence of a common root of these two languages, and to show that this root was grounded in a migration of people originating from the Caucasus. The Caucasus was considered the *Urheimat* (the original homeland) for the superior human race. From the *Urheimat* some groups might have migrated in the direction of Greece, while others continued in the direction of Germany. The original language from the *Urheimat* was referred to as *Indogermanish* (Indo-Germanic); later, this artificially invented language was called Indo-European.[4]

As a result of the two manoeuvres, Euclid's *Elements* could be distinguished as being the first presentation of genuine mathematics. We do not know much about Euclid except that he worked in Alexandria in Africa. He wrote in Greek, which was the universal academic language of the time. However, the European perspective ascribed Euclid with a unique role as being the first to present mathematics in the proper deductive format, and it presented Euclid as a European.

The School of Athens provides an exemplary illustration of the European colonisation of mathematics, as well as of philosophy. Euclid, as played by Donato Bramante, does not look like an African might look; he is a white Italian. The legend says that Pythagoras was born on the Greek island of Samos, but the Pythagoras portrayed in The School of Athens with his light reddish-brown hair and beard looks rather like a person born on the Faroe Islands. If we look for people with a darker skin colour, Ibn Rušd might be the only one to be found. Ibn Rušd is craning his neck in order to copy what Pythagoras is writing. The young people that are listening to what the "old white men" are saying all appear to be "young white men". One hardly ever finds a dark-haired person; Michelangelo's Heraclitus seems to be about the only one.

Rafael's School of Athens reflects the Eurocentrism of the Renaissance, as well as later versions of Eurocentrism, including the celebration of the "Caucasian" and "Aryan race". Rafael's School of Athens symbolises in an exemplary way the European colonisation of mathematics and philosophy.

---

[2]For a discussion of rigour in traditional Hindu mathematics, see Shrestha (2023).

[3]Bernal (1987) observes: "Since the 1840s Indo-European philology [. . .] has been at the heart of the Aryan Model. Then, as now, Indo-Europeanists and Greek philologists have been extraordinarily reluctant to see any connections between Greek—on the one hand—and Egyptian and Semitic, the two major non-Indo-European languages of the Ancient East Mediterranean, on the other" (p. 4).

[4]See Bernal (1987, p. 245).

Many presentations of the history of mathematics reiterate this Eurocentric view, and in *Mathematics in Western Culture*, Morris Kline (1953) gives a condensed formulation of it:

> Mathematics is a living plant which has flourished and languished with the rise and fall of civilizations. Created in some prehistoric period it struggled for existence through centuries of prehistory and further centuries of recorded history. It finally secured a firm grip on life in the highly congenial soil of Greece and waxed strong for a brief period. In this period it produced one perfect flower, Euclidean geometry. The buds of other flowers opened slightly and with close inspection the outlines of trigonometry and algebra could be discerned; but these flowers withered with the decline of Greek civilization, and the plant remained dormant for one thousand years. (p. 10)

Previously, I have referred to a remark made by William L. Schaaf (1930) about the significance of the concept of function.[5] He highlighted the concept as being a keynote of Western culture, which has no counterpart in any other culture. This remark can be seen as yet another expression of the colonisation of mathematics.

12.3 Modern Racism

The scientific revolution paved the way for the era of Modernity, and at the same time paved the way for scientific racism. We can see these two phenomena as being strictly independent, and this is normally what is taking place. While the scientific revolution has been celebrated as a breakthrough in human rationality, scientific racism has been considered a deplorable pseudo-scientific phenomenon. However, the two phenomena developed together as cohesive features of the modern worldview.

Scientific racism developed in an explicit form at the beginning of the nineteenth century, instigated by the growth of colonialism, slavery, and apparent genocides that were in need of "justification". Scientific racism developed as a university discipline, and tried to observe the quality criteria associated with respectable empirical studies. It was established at universities in England, France, Germany, and the USA, and it had an impact on many other scientific disciplines, including the history of mathematics.

Racism claims that there are qualitatively—and measurably—different human races, and that racial differences are rooted in biological differences, not in cultural diversities.[6] In the beginning of the modern era, in order for this claim to be substantiated, it needed to reinterpret the Bible's narrative of the book of Genesis. This was an urgent task as many groups with strong religious convictions were rapidly embracing racist positions, including many of the settlers in the USA. One alternative to the Bible story about Adam and Eve was the claim that God had

[5] See Sect. 9.3.

[6] See Sect. 1.2 for a clarification of what I mean by racism.

created not one, but different human races. This claim was made, for instance, by Charles Caldwell (1772–1853), who was a USA physician and member of the American Antiquarian Society. He postulated that God had created four human races: the Caucasian, the Mongolian, the American Indian, and the African, and that no transition between these races was possible.

The second feature of racist dogmatism is to claim that some human races are superior to other races. As such, human races were claimed to be organised in a hierarchy, with the Caucasian race in the top position. George Cuvier (1769–1832), a French zoologist dedicated to comparative anatomic studies, contributed to the further development of scientific racism. He was a highly esteemed scientist and a foreign member of the Royal Society of London for Improving Natural Knowledge, the Royal Swedish Academy of Sciences, and of the American Academy of Arts and Sciences. Cuvier claimed explicitly: "The white race [...] to which the civilized people of Europe belong [...] is superior to others" (Cuvier, quoted by Whitt & Clarke, 2019, pp. 56–57). "Caucasian" is a label invented as part of the racist worldview. It was introduced in 1785 by the German philosopher Christoph Meiners (1747–1810) to refer to a superior human race. The Bible tells us that Noah's Ark landed on Mount Ararat, located in the Caucasian Mountains, and the superiority of the stipulated Caucasian race could draw from this story. The German physician Johann Friedrich Blumenbach (1752–1840) engaged in systematic "racial classifications", applying the approach introduced by the Swedish biologist Carl Linnaeus (1707–1778). Blumenbach held the Caucasians up as the most beautiful and intelligent human race, while other races were considered degenerations from this ideal.[7]

In order to create an empirical basis for the claim about superiority, phrenology held a crucial position.[8] The assumption was that racial differences were rooted in biological structures, and that the human skeleton reflected such differences. The most important difference could be identified by studying the cranium, and craniometry developed into an esteemed scientific discipline. Samuel George Morton (1799–1851) from the USA contributed to the further development of scientific racism. In line with Caldwell, Morton found that the Bible could be interpreted according to a polygenetic view, meaning that God had created distinct human races. He believed that the races were created in a hierarchical order with the "white man" on top. Morton had a collection of more than 1000 skulls, many from the indigenous population in the USA. The skulls were carefully studied and measured in an attempt to prove his racist claim that the indigenous population in the USA belonged to an inferior race.

The hierarchy between races was claimed to establish a hierarchy between civilisations: superior civilisations were rooted in superior races. This conviction, however, caused difficulty for scientific racism. The construction of the Egyptian pyramids was acknowledged as a sublime achievement, but how could this be accomplished by an inferior civilisation? The attempted solution to this problem

[7] See Bernal (1987, pp. 237–238).

[8] The development of phrenology is referred to in Sect. 1.2.

was to show that the Egyptian civilisation, although located in Africa, was *not* an African civilisation. The study of the mummified pharaohs made it possible to claim that the shape of the cranium of the pharaohs were more like the Caucasian ideal than the shape of the cranium of black people.

The third feature of racist dogmatism is that superior races have rights that inferior races do not have. In particular, a superior race has the right to dominate an inferior race, and consequently the right to colonise inferior civilisations. In this way, the imperialist world order received its intellectual and political legitimation. In 1910 a version of this dogma was expressed in English Parliament when addressing the "Egyptian question". Egypt was colonised, but a range of administrative problems had emerged, so why not simply leave Egypt aside? Arthur James Balfour (1848–1930), who was a prominent conservative member of Parliament and who had been prime minister from 1902 to 1905, pointed out that as white Western nations demonstrate capacities for self-government, one never finds traces of self-government in Eastern countries. Balfour stated that among the "Orientals" one does not meet the capacity for self-government. As they cannot take properly care of themselves, it becomes urgent that somebody else assumes this responsibility. According to Balfour, this was "dirty work" but also "necessary labour". Balfour asked why the governing of inferior races should be the responsibility of the British, and he provided the answer: that people from the East, including Egypt, experienced better government by the British than they "ever had before". Balfour concluded: "We are in Egypt not merely for the sake of the Egyptian, though we are there for their sake; we are there also for the sake of Europe at large."[9] Consequently, Balfour made it appear that England not only had the right, but also the obligation to "take care of" an Eastern country such as Egypt. Imperialism had been turned into a generous humanitarian act.

A pitiless version of the third feature of racism is that superior races not only have the right to dominate inferior races, but also the right to eliminate them. This is the claim through which the North American Genocide was legitimised. This genocide became integrated into part of the Modern narrative about progress. The original population in the USA was characterised as "wild" and "ferocious". There was no possibility to establish a peaceful existence together, as the "wildness" and the "ferocity" were rooted in the biological structure of the indigenous population. Consequently, the progress of the modern project, taking the form of a move toward the West, had to be secured by a "war of extinction". In the Californian press of the time, one can find warnings about upcoming existential threats: "Extinction is no longer a question of time – the time has already arrived, the work has been commenced, and let the first white man who says treaty or peace be regarded as a traitor and coward" (quoted in Whitt and Clarke, 2019, p. 58).[10] Many of the massacres of the indigenous North American population were committed by the USA army, but there were also active groups of "Indian killers". These groups

[9]Extracts from Balfour's speech can be found in Said (1979, pp. 32–33).

[10]Madler (2016) documents how the genocide unfolded in California from 1846 to 1873.

definitely did not say "treaty" or "peace". They were not "traitors" nor "cowards", but according to the Californian press they were war heroes.

In magazines, newspapers and books from the time, one finds numerous illustrations of the theme: brave white men shooting ferocious and bloodthirsty Indians. In a drawing from the beginning of the 1860s, one sees a group of white men shooting towards an Indian village.[11] This is an explicit illustration of the genocide, although presented as a brave praiseworthy act. The North American Genocide has been glossed over, if not glorified, by such illustrations, and later by a vast canon of literature and films referred to as "Westerns". Many "Westerns" illustrate the pathological argument for genocide: some of them did horrible things to us, therefore we are justified in attacking all of them. The scientific racism at work here was not only part of the scientific worldview of the time, it also became a constituent of common-sense opinions.

In 1948 the newly founded United Nations approved a convention on the Prevention and Punishment of the Crime of Genocide, also referred to as the Genocide Convention. Here it was specified that genocide was an "act committed with intent to destroy, in whole or in part, a national, ethnical, racial or religious group as such".[12] Today we still witness genocidal acts, and we continue to hear the same pathological arguments expressed for genocide: some of them did horrible things to us, therefore we are justified in attacking all of them.

The European colonisation of mathematics was a crucial component for maintaining the dogma that superior achievement was accomplished by superior civilisations formed by superior groups of people. As with other forms of colonisation, the colonisation of mathematics is an integral component of Modernity, with devastating impacts on world orders and worldviews. This makes the decolonisation of mathematics critically important, but due to mathematics being embedded in Western thinking, such decolonisation is also extremely difficult.

12.4 Decolonisation of Mathematics

One issue in the decolonisation of mathematics concerns the concept of "Ancient Greek". As highlighted by Jens Høyrup (2020): "the ancient Greeks were not Europeans "(p. 7). Rather, the ancient Greek were connected to cultures around the Mediterranean, to the East, and to Africa. Martin Bernal (1987) had documented

[11] See Madler (2016, p. 278).

[12] This includes: "(a) Killing of members of the group. (b) Causing serious bodily or mental harms to the members of the group. (c) Deliberate inflicting on the groups conditions of life calculated to bring about its physical destruction in whole or in part. (d) Imposing measures intended to prevent births within the group. (e) Forcible transferring children of the group to another groups,". See the Convention for the Prevention and Punishment of the Crime of Genocide, reprinted in Whitt and Clake (2019, pp. 242–246).

the close relationship between the Greek language and other languages spoken around the Mediterranean.

The myth that Greek mathematics was European mathematics has been criticised severely in *The Crest of the Peacock: The non-European Roots of Mathematics*, where George Gheverghese Joseph (2000) outlines in detail the multicultural roots of mathematics. He shows the important contributions to mathematics coming from India, China, Japan, Africa, and South America. Mathematics does not have a taproot, but instead has a diversity of roots. Joseph makes it clear that the current predominant taproot presentation of the history of mathematics is false.

Another important contribution to the decolonisation of mathematics is found in the article "The formation of a myth: Greek mathematics – our mathematics", where Høyrup (1996) carefully analyses the fabrication of the Eurocentric myth that he ironically refers to as "conventional wisdom".[13] He observes:

> According to conventional wisdom, European mathematics [...] originated among the Greeks between the epochs of Thales and Euclid, was borrowed and well preserved by the Arabs in the early Middle Ages, and brought back to its authentic homeland by Europeans in the twelfth and thirteenth centuries [...]. Since then, it has pursued its career triumphantly. (p. 103)

Then he observes that this conventional wisdom is nothing but a carefully elaborated myth without any empirical substantiation:

> This tale, more or less biased or false as an historical account, has none the less become material truth in the sense that it has contributed to the self-understanding and thereby to the cultural identity of the European mathematical community/communities for centuries. (p. 103)

The false myth has materialised into an unquestionable given. It has been repeated numerous times; for instance, by Morris Kline (1953) in *Mathematics in Western Culture*, where he tells us: "The Egyptian and Babylonian civilizations drew inspirations for mathematical activities from many human needs and interests, but they fell short of greatness both in their understanding of mathematics and their contributions to the subject" (p. 21). For centuries this myth has contributed to the self-understanding of the European mathematical community, and today it is duplicated in a variety of mathematical textbooks.

The formation of the false myth took place in steps, and some major ones were taken during the Renaissance. Ancient Greece was celebrated as never before, and Renaissance humanists like Francesco Petrarca (1304–1374) contributed to this celebration. As Høyrup (1996) points out, although the majority of the Renaissance humanists knew relatively little about mathematics, they propagated the myth of an intimate relationship between European and Greek mathematics. However, even by the time of the Renaissance, this myth was questionable. For instance, Nicole Oresme (1320/1325–1382) who contributed to the development of mathematics,

[13] See also Greer (2021).

was well aware of the multiplicity of roots of mathematics and of the many original contributions to mathematics made by, for instance, Arabic scholars.

Algebra caused difficulty for the Eurocentric interpretation of mathematics. According to the myth, the role of the Arabic scholars was to ensure careful preservation of what already had been created in the Ancient Greece. But what about algebra? There does not seem to be a clear root of algebra in Greek mathematics. One "solution" to this problem was to consider algebra as being of less significance to mathematics compared to geometric thinking. Another solution to the algebra problem was to consider algebraic thinking as also being a Greek creation, although this was not expressed so clearly as in geometric thinking. Both solutions would eliminate the significance of the obvious Arabic contributions to mathematics. Both solutions appeared artificial, but nevertheless the Greek-Europe myth was maintained.

Petrus Ramus (1515–1572) was a French humanist, who was also well-oriented in mathematics. He outlined the history of mathematics, starting with the Bible's narrative about Adam and Eve. His principal point was that only the Patriarchs (Abraham, Isaac, and Jacob), the Greeks, and modern Europeans made significant contributions to mathematics. Having Rasmus' familiarity with mathematics in mind, Høyrup (1996) points out that Ramus was both dishonest and ignorant when he expressed the Eurocentric myth of mathematics.

A particular chapter in the colonisation of mathematics concerns India. Many examples have been presented that demonstrate that important elements of calculus were developed in India before similar ideas were presented by Leibnitz and Newton.[14] There is considerable evidence that shows that results from Indian mathematics found their ways to a European context. However, in order to preserve the Greek-Europe myth, the chapter about Indian mathematics had been torn out of the history books, as was the chapter about Arabic contributions to mathematics.

More recently, ethnomathematical studies have made huge contributions to the decolonisation of mathematics. In order to grasp the significance of these contributions, one needs to remember that the colonisation of mathematics was not the result of some researcher's preconceptions, but an expression of deep-rooted paradigmatic worldviews of the modern project. It is this project that ethnomathematical studies are confronting. Naturally, one finds ethnomathematical studies that address ethnomathematical practices as "picturesque" ethnographic phenomena, and recast ethnomathematics in the form of "Us" studying "Them". However, many ethnomathematical studies directly face the Eurocentric perspective on mathematics, such as *Ethnomathematics: Challenging Eurocentrism in mathematics education*, edited by Arthur B. Powell and Marilyn Frankenstein (1997).[15] Instead of thinking of ethnomathematics as a study of different enculturations of mathematics, one can think of it as a study of relationships between such enculturations. This idea has been

[14] See Almeida and Joseph (2009) and Joseph (2009).

[15] For an overview of ethnomathematical studies today, see Rosa et al. (2016), and Borba and Orey (2023).

presented by Aldo Parra (2018, in progress). By seeing ethnomathematics as a study of relationships, the temptation to see it as a study of "picturesque" phenomena is eliminated. This is crucial for undertaking the decolonisation of mathematics.

In the article "Bringing an Ethnomathematical Perspective into Classrooms", Swapna Mukhopadhya and Brian Greer (in progress) offer the following observation: "A central thrust in the inherently political nature of Ethnomathematics has always been the construction of a counternarrative to the Eurocentric myth about the development of academic mathematics, a myth that can be seen as a long-established, deeply-entrenched, and ongoing manifestation of White (intellectual) supremacy." I agree completely with this observation: ethnomathematics can be a powerful way of decolonising mathematics.[16]

Over time architecture has found a huge variety of expressions by way of monumental constructions such as the Mayan pyramids, the Egyptian pyramids, Stonehenge, the Great Wall of China, the Parthenon, the Pantheon, St Peter's Basilica, the Taj Mahal Mausoleum, and the Kremlin. One can also think of igloos, the white painted Greek cities clinging to rocky sides of small island in the Aegean Sea, the wooden houses of the Tyrolean country side, the apartment blocks for workers during the period when Manchester became known as Cottonopolis, the merchant homes in Amsterdam built when the Netherlands first developed as a colonial power, the sand-coloured houses in historic Timbuktu, and the tents of the indigenous population of North America. Architecture has never assumed any definite format. Nevertheless, in architecture one also meets versions of the white-supremacy ideology, such as in the list of the most famous architects of all time, which includes only architects from First World Countries.[17] However, like any human construction, both architecture and mathematics are indefinite with respect to culture. No culture can claim to engender the definite architecture, whether of buildings or of mathematics.

The purpose of Part III of this book was to acknowledge that mathematics is a human construction which does not exist in an absolute form, but as always undergoing development. Mathematics is formed through indefinite constructions.

For a critical philosophy of mathematics, it is important to grasp mathematics as being indefinite with respect to concepts, proofs, topics, applications, power, and culture. Mathematical constructions are temporary, and can always be reworked, further elaborated, or maybe abandoned. Mathematics does not exist in a definite

---

[16]In Skovsmose (in progress), I explain why I am reluctant to use the expression "Western mathematics", although this is often used by people who highlight the importance of ethnomathematical studies. My reason is that, if we look around the world today, there is nothing Western about mathematics. Mathematics research takes place all over the world; mathematics is applied everywhere; and schools everywhere have mathematics as a discipline. If we go back in history, mathematics was not "Western". The "Westernisation" of mathematics is an ideological construction.

[17]One such list includes only two from Spain, two from USA, one from Germany, one from Finland/USA, two from England, one from Canada/USA, and one from Italy. All men, and all from First World Countries. See https: 17 Most Famous Architects of All Time (timeout.com).

format, and there is nothing to be labelled "genuine mathematics". To a critical philosophy of mathematics, it is important not to be trapped by Eurocentrism, but to recognise the multitude of enculturations of mathematics. This makes a critique of mathematics not only possible, but necessary.

References

Almeida, D. F., & Joseph, G. G. (2009). Kerala mathematics and its possible transmission to Europe. In P. Ernest, B. Greer, & B. Sriraman (Eds.), *Critical issues in mathematics education* (pp. 171–188). Information Age Publishing.

Bernal, M. (1987). *Black Athena: Afroasiatic roots of classical civilization. Volume 1: The fabrication of ancient Greece 1785–1985*. Rutgers University Press.

Borba, M. C., & Orey, D. C. (Eds.). (2023). *Ubiratan D'Ambrosio and mathematics education: Trajectory, legacy and future*. Springer.

Greer, B. (2021). Learning from history: Jens Høyrup on mathematics, education, and society. In D. Kollosche (Ed.), *Exploring new ways to connect: Proceedings of the eleventh international mathematics education and society conference, volume. 2* (pp. 487–496). Tredition.

Høyrup, J. (1996). The formation of a myth: Greek mathematics – Our mathematics. In C. Goldstein, J. Gray, & J. Rilter (Eds.), *L'Europe mathématique, mathematical Europe* (pp. 103–119). Editions de la Maison des sciences de l'homme.

Høyrup, J. (2020). *From Hesiod to Saussure, from Hippocrates to Jevons: An introduction to the history of scientific thought between Iran and the Atlantic*. Preprint. Max Planck Institute for the History of Science.

Joseph, G. G. (2000). *The crest of the peacock: The non-European roots of mathematics*. Princeton University Press.

Joseph, G. G. (2009). *Kerala mathematics: History and its possible transmission to Europe*. Kerala Council for Historical Research and Art and Humanities Research Board.

Kline, M. (1953). *Mathematics in Western culture*. Oxford University Press.

Madler, B. (2016). *An American genocide: The United States and the California Indian catastrophe, 1846–1873*. Yale University Press.

Mukhopadhya, S., & Greer, B. (in progress). Bringing an ethnomathematical perspective into classrooms. In B. Greer, D. Kollosche, & O. Skovsmose (Eds.), *Breaking images: Iconoclastic analyses of mathematics and its education*. Open Book Publishers.

Parra, A. I. S. (2018). *Curupira's walk: Prowling ethnomathematics theory through decoloniality*. Doctoral dissertation, Aalborg University Press.

Parra, A. (in progress). A performative and relational ethnomathematics. In B. Greer, D. Kollosche, & O. Skovsmose (Eds.), *Breaking images: Iconoclastic analyses of mathematics and its education*. Open Book Publishers.

Powell, A. B., & Frankenstein, M. (1997). *Ethnomathematics: Challenging eurocentrism in mathematics education*. SUNY Press.

Rosa, M., D'Ambrosio, U., Orey, D. C., Shirley, L., Alangui, W. V., Palhares, P., & Gavarrete, M. E. (Eds.). (2016). *Current and future perspectives of ethnomathematics as a program*. Springer.

Said, E. (1979). *Orientalism*. Vintage books.

Schaaf, W. L. (1930). Mathematics and world history. *The Mathematics Teacher, 23*(8), 496–503.

Shrestha, M. B. (2023). Philosophy, rigor, and axiomatics in mathematics imposed or intimately related? In M. A. V. Bicudo, B. Czarnocha, M. Rosa, & M. Marciniak (Eds.), *Ongoing advancements in philosophy of mathematics education* (pp. 409–428). Springer.

Skovsmose, O. (in progress). A critical conception of mathematics. In B. Greer, D. Kollosche, & O. Skovsmose (Eds.), *Breaking images: Iconoclastic analyses of mathematics and its education*. Open Book Publishers.

Whitt, L., & Clarke, A. W. (2019). *North American genocides: Indigenous nations, settler colonialism, and international law*. Cambridge University Press.

Part IV
Ethical Challenges

Chapter 13
A Performative Interpretation of Mathematics

Abstract My performative interpretation of mathematics is inspired by a performative interpretation of language. I elaborate on this interpretation by discussing language as description, mathematics as description, language as performance, and mathematics as performance. The descriptive interpretation of language reflects a conception of reality as being "out there" waiting to be depicted. This interpretation has been expressed by the picture theory of language turning into a descriptive interpretation of mathematics, as formulated by Wittgenstein. Originally, the performative interpretation of language was elaborated within analytic philosophy, but it developed profoundly through discourse theory. A discourse forms our experiences, knowledge, ideas, assumptions, preconceptions, superstitions, and ideologies. It forms categories for understanding as well as for misunderstanding. Like discourse, mathematics forms our experiences, knowledge, ideas, assumptions, preconceptions, superstitions, and ideologies. It also forms categories for understanding and misunderstanding. Mathematics forms our worldviews, and also our life-worlds.

Keywords Descriptive interpretation of language · Descriptive interpretation of mathematics · Performative interpretation of language · Performative interpretation of mathematics · Romantic movement · Linguistic relativism · Analytic philosophy · Speech act · Discourse theory · Discourse · Picture theory · Worldview · Life-world

Mathematics has long been considered a means of description. As such, it can portray natural, social, indeed any kinds of phenomena. Such a *descriptive interpretation* of mathematics has led mathematics to be called the language of science. This interpretation has also dominated the discussion of mathematical modelling, including that which is taking place in mathematics education.

I find the descriptive interpretation of mathematics to be inadequate, and inspired by a performative interpretation of language, I am going to suggest a *performative interpretation* of mathematics. This brings into focus what is done, or performed, by means of mathematics. Mathematical descriptions bring about interventions in what we assume we are describing, by selecting, highlighting, ignoring, and forgetting. Through doing mathematics, we format what we are describing and consequently how we are acting.

O. Skovsmose, *Critical Philosophy of Mathematics*, Advances in Mathematics Education, https://doi.org/10.1007/978-3-031-71375-0_13

The performative interpretation of mathematics exposes the intimate connection between mathematics and action. Any action, including mathematics-based action, could have various characteristics and consequences. They could be useful, risky, dangerous, expensive, profitable, benevolent, or horrendous. All actions need to be critically addressed, and consequently, the performative interpretation of mathematics leads to an ethical challenge. In this context I see ethics as a broad discussion of the socio-political impact of mathematical performatives, and consequently ethics becomes an important dimension of a critical philosophy of mathematics. Here, in Part IV, I deal with the ethical challenges that are faced by mathematical practices in research, application, and education.

In *Towards of Philosophy of Critical Mathematics Education* (Skovsmose, 1994), I introduced the notion of the formatting power of mathematics, by means of which I expressed a preliminary version of a performative interpretation of mathematics. This interpretation was developed further by investigating features of bringing mathematics into action (see, for instance, Skovsmose, 2005). In "A Performative Interpretation of Mathematics" (Skovsmose, 2023), I give an explicit formulation of this interpretation, which I am going to draw on in this chapter.[1] Here I will discuss language as description, mathematics as description, language as performance, and mathematics as performance.

13.1 Language as Description

Language has been considered a tool for describing our observations and formulating our opinions. When we achieve mastery of language, this tool is at our disposition.

The empiricists John Locke, George Berkeley, and David Hume never questioned whether the language at their disposition was adequate for formulating their views. It was considered a transparent means for expressing what they wanted to describe. When Immanuel Kant formulated a critique of reason by identifying some basic epistemic categories, he never questioned whether the language that he had at hand was adequate for expressing such a critique. Language as such, in and of itself, was not considered to cause any philosophical problems. Rather, it was a means for solving such problems.

In *Confessions*, written between 397 and 400, Augustine (1961) made a comment about how he learned language, where he expressed a descriptive interpretation of language:

> I noticed that people would name some object and then turn towards whatever it was they had named. I watched them and understood that the sound they made when the wanted to indicate that particular thing was the name which they gave to it, and their actions clearly

---

[1] A performative interpretation of mathematics was elaborated upon in Ravn and Skovsmose (2019). See also Ernest (2021) and Skovsmose (2023).

> showed what the meant [. . .] So by hearing words arranged in various phrases and constantly repeated, I gradually pieced together what they stood . . . (Book I, §8, p. 29).

The descriptive interpretation of language reflects a conception of reality as being "out there", waiting for us to describe it. This interpretation operates with a clear distinction between what is being described and the description itself. It operates with a sharp subject-object distinction, and not with the possibility of any dialectic between language and reality. Language does not interact any more with reality than a mirror interacts with what it is mirroring.

As part of the so-called linguistic turn in philosophy, launched by Gottlob Frege and Bertrand Russell, language eventually became recognised as a philosophic problem. It appeared that natural language had many in-built limitations. Analytic philosophy pointed out that the grammar of natural language was too loose, leaving space for ambiguities, contradictions, and misunderstandings to occur. Natural language fell under suspicion, which was expressed by Russell (1905) in the paper "On Denoting". He found that a way out of the difficulties was to start operating with formal languages. They had an analytical power in depicting facts in a way that no natural language could demonstrate. Formal languages could expose an underlying logic, which remained concealed by the grammar of the natural language.

13.2 Mathematics as Description

In the *Tractatus Logico-Philosophicus*, published in a German-English parallel edition in 1922, Ludwig Wittgenstein (1922) presented a picture theory of language. This is an explicit expression of a descriptive interpretation of language as well as of mathematics.[2]

During the First World War, Wittgenstein was enlisted as a soldier in the Austrian army, and he was stationed at the Eastern front. During the whole war period he continued working on the manuscript of the *Tractatus*, and during this time he added the picture theory of language to the manuscript. During the final months of the war Wittgenstein was called on to serve at the Southern front. In 1918, when the Austrian army in Italy surrendered, Wittgenstein was taken as a prisoner of war and stayed in a concentration camp until 1919. At that time the *Tractatus* was in its final form and was prepared for publication.

Wittgenstein's picture theory includes two components: the ontological component concerns the structure of reality, while the epistemic component concerns the structure of the language, which should be able to capture reality.

Wittgenstein started the *Tractatus* by making an audacious ontological claim: the world is composed of facts and not of things. By making this claim Wittgenstein

---

[2] In 1921, the *Tractatus* was published in a German journal, but this publication included so many typographical mistakes that Wittgenstein only considered the 1922 version as being an actual publication of his work.

distanced himself from a long philosophic tradition of seeing the world as being composed of some kind of objects. The Pythagoreans saw the world as composed of numbers, Democritus saw it as composed of atoms, Plato as composed of ideas, Galilei as composed of material objects defined by their shape, size, and moments, and Leibnitz as composed of monads. Wittgenstein was well aware of the power of the opening statements of the *Tractatus*: "The world is everything that is the case. The world is the totality of facts, not of things" (§1 and §1.1). One might even come to think of the opening statement of the Gospel of John: "In the beginning was the Word, and the Word was with God, and the Word was God" (John 1: 1).

The epistemic claim made by Wittgenstein is that facts can be pictured by language. It has been reported that in 1914 Wittgenstein got his inspiration for the theory by reading about a lawsuit concerning a traffic accident. In order to clarify for the judge what had taken place, toy cars were brought to the courtroom, by means of which the accident was reconstructed. The real accident was depicted by means of the toy cars. In an analogical way, Wittgenstein saw language as picturing real states of affairs.[3]

In order for a language to picture reality, it needs to have something in common with reality. Wittgenstein referred to this common thing as a "logical form". He stated that the picture of a fact must have the same logical form as the fact itself (§2.2). Language and reality must share logical forms. As reality is composed of facts, some atomic and some molecular, the language that pictures reality must be composed of statements, some atomic and some molecular. An atomic statement is true when it pictures an atomic fact, and a molecular statement is true when it pictures a molecular fact. The logical form of a formal language as presented in *Principia Mathematica* by Alfred Whitehead and Bertrand Russell (1910, 1913a, 1913b) is equivalent to the logical form of reality. This language is truth functional, by which is meant that the truth value of a molecular statement is fully determined by the truth values of the atomic statements of which it is composed. The logical forms of a truth-functional language correspond to the logical form of the world, and this makes picturing possible. If we assume a logicist position, seeing logic and mathematics as integrated phenomena, Wittgenstein's picture theory of language turns into a picture theory of mathematics: by means of mathematics one is able to picture reality.[4]

Formalism presents a descriptive interpretation of mathematics. In *Outlines of a Formalist Philosophy of Mathematics*, Haskell Curry (1951) drew a distinction between two notions of truth, both related to mathematics. One can talk about the possible truth of a mathematical statement, but also about the possible truth of a

---

[3]In his biographical sketch Wright (1955) told the story of Wittgenstein's possible inspiration for the picture theory of language. Klagge (2020), making comments about each and every paragraph of the *Tractatus*, mentions that it has not been possible to identify which article Wittgenstein had been reading. However, Klagge shows an illustrated page from the magazine *Popular Mechanics* from June 1914, with the headline "Lawyer Studies Accident with Plat and Toys" (p. 68).

[4]For a presentation of the development of Wittgenstein's conception of mathematics, see Floyd (2005).

mathematical theory. He highlighted the importance of not mixing up these two notions of truth.

According to Curry, the truth of a mathematical statement is always relative to a formal system; there are no absolutely true mathematical statements. Statements like $2 + 2 = 4$, the sum of the angles of a triangle is 180°, and a group has one and only one neutral element are not absolute truths. The proposition $2 + 2 = 4$ is true if we assume Peano's axiomatisation of the natural numbers; the statement that the sum of the angles in a triangle is 180° is true if we assume a Euclidian axiomatisation of plane geometry; and the theorem that a group has one and only one identity element is true if we assume the axioms of group theory.

The situation is quite different when we consider the possible truth of a mathematical theory. In this case we are dealing with a notion of truth similar to the one applied in science in general. The truth of a mathematical *theory* is an empirical issue, while the truth of a mathematical *statement* is a logical issue. If a mathematical theory can be applied successfully with respect to a set of empirical phenomena, the theory might be considered true. The theory of natural numbers turns out to be true when it is applied successfully in everyday situations. Euclidean geometry is true to the extent that it can be applied successfully with respect to, for instance, making architectural designs. Naturally, there are situations where Euclidian geometry turns out to be inadequate, such as when we are dealing with spatial phenomena, when the theory of relativity turns out to be more appropriate. The group theory axioms can be successful when we are dealing with operations such as reflecting and rotating. The way Curry addressed the possible truth of a mathematical theory is in accordance with a descriptive interpretation of mathematics. In line with Wittgenstein's picture theory, a description can be accurate and useful, and in this sense it can be true. This is an empirical question.

Within mathematics education, mathematical modelling has been discussed quite broadly. Many of these discussions, also assuming a version of Wittgenstein's picture theory, consider a mathematical model as being a more or less accurate picture of reality. The part of reality that is going to be modelled is described in empirical or perceptual terms, and so are the problems that are going to be addressed. The empirical description of reality is then translated into a mathematical vocabulary where empirical notions and relationships are correlated with mathematical entities, and the real-life problem is correlated with a mathematical problem. The mathematical representation of the real-life situation—the mathematical model—is then investigated with mathematical means, and a solution to the mathematical problem might be identified. This mathematical solution, then, becomes reinterpreted in empirical terms and is assumed to be a solution to the original real-life problem. It is also assumed that the adjustment of the model to reality needs to go through several loops in order to obtain accuracy.[5]

[5] The descriptive interpretation of mathematical modelling has been expressed in what can be called the canonical representation of mathematical modelling, see Ravn and Skovsmose (2019, pp. 152–153).

I find this descriptive conception of mathematical modelling to be entirely misleading, but unfortunately it is common in mathematics education literature.[6] I find it extremely important to recognise that mathematics is not merely a tool for depicting reality, but that mathematics itself might bring about a thorough reconfiguration of that reality.

13.3 Language as Performance

The performative interpretation of language has several roots, and I will refer to three of them. One is found in the Romantic Movement in Germany, a second in the study of indigenous languages, and a third in analytic philosophy.

The Romantic Movement in Germany did not assume that language was a transparent tool for making descriptions, but saw it as a deep-rooted cultural phenomenon. Language was an expression of the *Volkgeist*. The German word *Volk* means "people", and *Geist* means "spirit", so *Volkgeist* can be read as the "spirit of the people". The German word *Vaterland* means fatherland and *Muttersprache* means mother tongue, and the Romantic Movement saw close relationships between the *Vaterland* and the *Muttersprache*. The fatherland established a natural environment for the development of the mother tongue. The Romantic Movement showed an intensive interest in language, which was seen as an expression of the *Volkgeist*. The brothers Jacob Grimm (1785–1863) and Wilhelm Grimm (1786–1859) were deeply interested in capturing the German *Volkgeist* as expressed in folklore. The result was *Grimms' Fairy Tales*. The Romantic Movement anticipated a performative interpretation of language by seeing language as formed by a *Volkgeist* and, simultaneously, as forming a *Volkgeist*.

With reference to a quite different context, Edward Sapir (1884–1939) and Benjamin Lee Whorf (1897–1941) presented a performative interpretation of language. Together they studied different native American languages, and the study of the Hopi language brought them to recognise that language operates as much more than a descriptive device. Language shapes how people perceive and conceptualise the world. A language forms worldviews, and in doing so it shapes the world itself. Sapir (1929) condensed this idea in the following way:

> Human beings do not live in the objective world alone, nor alone in the world of social activity as ordinarily understood, but are very much at the mercy of the particular language which has become the medium of expression for their society. It is quite an illusion to image that one adjusts to reality essentially without the use of language and the language is merely

[6] See, for instance, Arseven (2015); Blum (1993); Blum et al. (2007); Blum and Niss (1991); Kaiser (2020); Kaiser and Sriraman (2006); de Lange et al. (1993); Niss (1989); Niss et al. (1991); and Wess et al. (2021). The common conception of mathematical modelling has been critically investigated by Schürmann (2023) as a contribution to the philosophy of mathematics educaion. See also Schürmann (2018) for a problematisation of the separation between mathematics and reality.

> an incidental means for solving specific problems of communication or reflection. The fact of the matter is that the "real world" is to a large extent built up on the language habit of the group. (p. 209)

Sapir is claiming that there is no "real world" as such; it is shaped by language. We are at the mercy of the language that we are subjected to. Sapir demolished the distinction between language and reality, by assuming the existence of a subject-object dialectic in the form of a language-reality dialectic.

Whorf (1956) added that language "is not merely a reproducing instrument for voicing ideas but rather is itself the shaper of ideas, the program and guide for the individual's mental activity" (p. 212). This means that our way of seeing the world becomes "part of a particular grammar" (p. 212). This brings about linguistic relativism, which Whorf expressed by stating that different grammars point toward "different types of observations and different evaluations of externally similar acts of observation" (p. 221). The users of different languages are located within different worldviews. In *Philosophical Investigations*, Wittgenstein distanced himself from the views of language that he expressed in the *Tractatus*. He abandoned the idea that there exists one and only one language that can picture reality, for assuming a linguistic relativism.

Within the outlook of analytic philosophy, both Ludwig Wittgenstein and John Austin took huge steps that brought them to a performative interpretation of language. Wittgenstein (1953) introduced the notion of language game, and Austin (1962) talked about "doing things with words". In his initial formulations, Austin seemed to suggest that some linguistic expressions—like promising, lying, and baptising—include a performative dimension, but in later formulations he highlighted that all kinds of linguistic expression are performative. Hence, we are dealing with an inherent feature of language. The notion of speech act, as developed by John Searle (1969) within the same frame of analytic philosophy, captures precisely the close relationship between language and action.

Discourse theory brings the performative interpretation of language beyond the boundaries set by the Romantic Movement, by Sapir and Whorf, and by analytic philosophy. The romantic understanding of language incorporated a range of nationalist conceptions; Sapir and Whorf's language relativism was closely associated with the study of indigenous languages; and the analytic philosophy of language disregarded the political roles played by language. Discourse theory investigates how discursive structures can be the carrier of political interest, of economic priorities, of racist and sexist positions, of preconceptions of any kind. Discourse theory helps us to see how discourses may form our worldviews, our decisions, our actions, and eventually our life-worlds.

In 1936, Edmund Husserl introduced the notion of life-world, when he published the original German version of *The Crisis of European Sciences and Transcendental Phenomenology* (see Husserl, 1970). The notion has been elaborated further within the phenomenological outlook and by social theorising inspired by this outlook. Husserl considered a life-world as being an original experienced world, not formed by scientific or any other kind of knowledge. My use of the notion of life-world is different to that of Husserl. I see a life-world as a complexity of life conditions structured by economic, political, religious, ideological, cultural, and discursive factors.[7]

Discourses have been described in a variety of ways, and let me briefly refer to some of them. *In Discourse Analysis*, Gillian Brown and George Yule (1983) pointed out that the analysis of discourse is "the analysis of language in use" (p. 1), which means that it cannot be restricted to the description of grammatical structures. In line with this they highlight that discourse analysis is committed to an investigation of "what the language is used for" (p. 1). In *Problems in General Linguistics*, Émily Benveniste (1971) also made it clear that by operating with the notion of discourse "we leave the domain of language as a system of signs and enter into another universe, that of language as an instrument of communication" (p. 110). However, while speech act theory focused on specific actions and on what a person might be doing with their speech, discourse theory looks much broader on linguistic performatives.

In *The Archaeology of Knowledge*, the first French edition of which appeared in 1969, Michel Foucault (1989) made broad use of the notion of discourse. According to him, a discourse is a symbolic system composed of conceptions, beliefs, attitudes, and dispositions for actions. A discourse itself is formed by a range of economic, political, and ideological factors. A variety of discourses with different historical formations exist—medical, psychological, psychiatric, natural scientific, and everyday—all of them loaded with assumptions and presumptions. A discourse produces what one might take for granted and consider to be true.

We might try to compare a discourse with a paradigm as described by Thomas Kuhn (1962), since a paradigm is also formed by a set of taken-for-granted assumptions. Nevertheless, I see Foucault's conception of discourse as being rather different. While Kuhn's conception of paradigm concerns scientific knowledge, Foucault's conception of discourse concerns the formation of any kind of knowledge. While Kuhn described a paradigm as a rather coherent system where ideas, assumptions, and perspectives have found an epistemic equilibrium, Foucault opened up the possibility that a discourse encompasses tensions and contradictions. And while Kuhn described a change of paradigm as being revolutionary, Foucault presented discursive changes as ongoing processes. A discourse is a non-transparent shaper of our views and actions. It sets the epistemic horizon, within which we are operating. Power is exercised through discourses by setting this horizon and, therefore, the agenda for what is said, and what is left unsaid.

In his discussions of the relationship between knowledge and power, Foucault looked at psychology, psychiatry, and conceptions of sexuality; he did not pay much attention to the natural sciences, nor to mathematics. However, I find that mathematics brought into action in an exemplary way demonstrates the close relationship between knowledge and power.

In their Introduction to *Colonial Discourse and Post-Colonial Theory*, Patrick Williams and Laura Chrisman (1994) outlined the complexity of colonial discourses

---

[7]For a clarification of my interpretation of life-world, see Skovsmose (2014).

and also of the many ways that they can be deconstructed. Colonial discourses are formed as ways of hiding the brutality of colonisations and of legitimising them in general. William and Chrisman referred to *Orientalism* by Edward Said (1979), who came from Palestine. By "orientalism" Said referred to the Western image of people from the East as being inferior and unable to rule themselves. This is a discursive act by means of which colonising the East turns into a humanitarian act of helping others.[8]

In *New Theories of Discourse*, Jacob Torfing (1999), drawing on a range of references, reached the following hard-boiled definition: "A discourse is a differential ensemble of signifying sequence in which meaning is constantly renegotiated" (p. 85). This definition can be unfolded in different ways, and it is revealed that the clarification of the notion of discourse is far from being only an analytical task; it is part of a political struggle.

Having such different observations in mind, let me suggest the following general characteristic of a discourse:

> A discourse is a contingent construction, which is always under reconstruction. It is an all-embracing epistemic system that structures experiences, attitudes, decisions, and actions. It need not be coherent, but may include contradictions and conflicting positions.
>
> A discourse captures human life in a fabric of ideas, assumptions, preconceptions, superstitions, and ideologies. It forms categories for understanding as well as for misunderstanding. However, we do not find any solid position from where to identify the nature of the discursive formation of categories. A discourse is a truth-fabricating system, although not in the sense that truth refers to any objective phenomena, but rather to discursive generated convictions.
>
> A discourse forms not only our worldviews, but also our life-worlds. It is an all-embracing ontological system. Discourses are manifestations of power, which can embody any kind of interests. Discourses do not operate in transparent ways. They can mask power, and make it operate incognito.

This summary of what a discourse could include captures what a performative interpretation of language might mean.

13.4 Mathematics as Performance

An important inspiration for a performative interpretation of mathematics comes from seeing mathematics as a discourse. This helps us to overcome the narrow perspective established by a descriptive interpretation of mathematics, which creates a depoliticised view of mathematics. Such a view has been expressed by logicism,

---

[8]In Sect. 12.3, I discussed the phenomenon of "Modern racism", which embraces the racist dogmatism stipulating some races to be inferior. Modern racism can be interpreted as a discourse that determines worldviews and consequently also world orders.

formalism, intuitionism, and the philosophy of mathematical practice. However, for the good and the bad, mathematics is involved in a variety of social phenomena. Seeing mathematics as a discourse is a way of recognising the socio-political complexities of mathematical performatives.

The modern philosophies of mathematics have, to a great extent, ignored technological issues, and the philosophy of technology has ignored mathematical issues. I find, however, that the philosophy of mathematics and the philosophy of technology are broadly overlapping and in need of each other. Technology has been interpreted as a tool at hand for human purposes, but technology has also been interpreted as an all-embracing structuring mechanism of all aspects of social life.[9] This second interpretation highlights that technology not only forms our material living conditions, but in addition our thinking and ways of handling problems. Technology forms all aspects of our life-worlds, as does mathematics.

I am not claiming that mathematics is a discourse, nor that mathematics is a technology. But I find inspiration for a performative interpretation of mathematics by tentatively seeing mathematics as a discursive and technological practice, and as an all-embracing structure.

> Mathematics is a contingent construction, which is always under reconstruction. Mathematics is indefinite with respect to concepts, proofs, topics, applications, power, and culture. Mathematics is an all-embracing epistemic system that structures experiences, attitudes, decisions, and actions. Insight provided by mathematics need not be coherent, but might include contradictions and conflicting positions.
>
> The nature of mathematical structuring of the reality within which we are submerged, is not easy to identify. We are engulfed in a fabric of ideas, assumptions, preconceptions, superstitions, and ideologies. Mathematics forms categories for understanding as well as for misunderstanding. However, we do not find any solid position from where to discuss the nature of such mathematical structurings. Mathematics is a truth-fabricating system, although not in the sense that truth refers to any objective phenomena, but rather to mathematical generated convictions.
>
> Mathematics forms epistemic structures as well as material realities. It is an all-embracing ontological system. It forms not only our worldviews, but also our life-worlds. Mathematical structurings are manifestations of power, which can embody any kind of interests. Mathematics brought into action does not operate in transparent ways. Mathematics masks power, and make it operate incognito.

By comparing mathematics with discourse or with technology, I do not try to justify a performative interpretation of mathematics. This needs exemplification of mathematical performatives, which will come in the following Chap. 14.

As mentioned previously, Wittgenstein made a bold metaphysical claim in the *Tractatus* when stating that the world consists of *facts*. By doing so, he opposed the

[9] See Sect. 3.3.

common empiricist view that the world is composed of things. If I should try to make a similar grand claim, it would be that the world is composed of *processes*, rather than of things or of facts. Communication, in whatever form it might have, is a process without termination. Production involves ongoing processes, including profit making, controlling, and exploitation. The stock market is a nucleus of ongoing economic transactions. International crises do emerge, develop, and become threatening. At times solutions are identified; much of the time no solution is in sight. Daily life is made up of processes of shopping, talking, cooking, watching TV, arguing, resting, and sleeping.

We are involved in processes whose outcomes are impossible to predict. We are subjected to processes many of which are out of our control, and perhaps out of human control completely.

When we consider mathematical performatives, I suggest we pay particular attention to how mathematics forms processes that make up our life-worlds.[10] In Chap. 14, I will show the way mathematics is shaping worldviews, fabricating the algorithmic workplace, sanctifying norms, splitting practices, forming the future, disguising power, and articulating social justice. The notions "shaping", "fabricating", "sanctifying", "splitting", "forming", "disguising", and "articulating" signal performances. Mathematical performatives concern processes rather than things or facts.

Mathematics is formed to perform. Strong interests are associated with the formation of mathematics, as well as with mathematical performatives. This does not mean that mathematical performatives are under control. Performatives that nobody had thought of in advance might occur. Mathematical performatives might be both powerful and obscure.

References

Arseven, A. (2015). Mathematical modelling approach in mathematics education. *Universal Journal of Educational Research, 3*(12), 973–980.

Augustine. (1961). *Confessions*. Penguin Books.

Austin, J. L. (1962). *How to do things with words*. Oxford University Press.

Benveniste, E. (1971). *Problems in general linguistics*. University of Miami Press.

Blum, W. (1993). Mathematical modelling in mathematics education and instruction. In Breiteig (Ed.), *Teaching and learning mathematics in context* (pp. 3–14). Ellis Horwood.

Blum, W., & Niss, M. (1991). Applied mathematical problem solving, modelling applications, and link to other subjects: State, trends and issues in mathematics education. *Educational Studies in Mathematics, 22*(1), 37–68.

Blum, W., Galbraith, P. L., Henn, H.-W., & Niss, M. (2007). *Modelling in applications in mathematics education*. Springer.

Brown, G., & Yule, G. (1983). *Discourse analysis*. Cambridge University Press.

Coeckelbergh, M. (2022). *Digital technologies, temporality, and the politics of co-existence*. Palgrave Macmillan.

[10] In his discussion of digital technologies, Coeckelbergh (2022) pays special attention to the notion of process.

Curry, H. B. (1951). *Outlines of a formalist philosophy of mathematics*. North-Holland Publishing Company.
Ernest, P. (2021). The ethics of mathematical practice. In B. Sriraman (Ed.), *Handbook of the history and philosophy of mathematical practice*. Springer.
Floyd, J. (2005). Wittgenstein on philosophy of logic and mathematics. In S. Shapiro (Ed.), *The Oxford handbook of philosophy of mathematics and logic* (pp. 75–128). Oxford University Press.
Foucault, M. (1989). *The archaeology of knowledge*. Routledge.
Husserl, E. (1970). *The crisis of European sciences and transcendental phenomenology*. Translated with an introduction by David Car. Northwestern University Press.
Kaiser, G. (2020). Mathematical modelling and applications in education. In S. Lerman (Ed.), *Encyclopedia of mathematics education* (2nd ed., pp. 553–561). Springer.
Kaiser, G., & Sriraman, B. (2006). A global survey of international perspectives on modelling in mathematics education. *ZDM, 38*(3), 302–310.
Klagge, J. C. (2020). *Tractatus in context: The essential background for appreciating Wittgenstein's Tractatus Logico-Philosophicus*. Routledge.
Kuhn, T. (1962). *The structure of scientific revolutions*. Chicago University Press.
Lange, J. de, Huntley, I., Keitel, C., & Niss, M. (Eds.). (1993). *Innovation in math education by modelling and applications*. Ellis Horwood.
Niss, M. (1989). Aims and scope of applications and modelling in mathematics curricula. In W. Blum et al. (Eds.), *Applications and modelling in learning and teaching mathematics* (pp. 22–31). Ellis Horwood.
Niss, M., Blum, W., & Huntley, I. (Eds.). (1991). *Teaching of mathematical modelling and applications*. Ellis Horwood.
Ravn, O., & Skovsmose, O. (2019). *Connecting humans to equations: A reinterpretation of the philosophy of mathematics*. Springer.
Russell, B. (1905). On denoting. *Mind, 14*(56), 479–493.
Said, E. (1979). *Orientalism*. Vintage Books.
Sapir, E. (1929). The status of linguistic as a science. *Language, 5*(4), 207–214.
Schürmann, U. (2018). The separation of mathematics from reality in scientific and educational discourse. In P. Ernest (Ed.), *The philosophy of mathematics education today* (pp. 241–251). Springer.
Schürmann, U. (2023). Mathematical modelling: A philosophy of science perspective. In M. A. V. Bicudo, B. Czarnocha, M. Rosa, & M. Marciniak (Eds.), *Ongoing advancements in philosophy of mathematics education* (pp. 309–329). Springer.
Searle, J. (1969). *Speech acts*. Cambridge University Press.
Skovsmose, O. (1994). *Towards a philosophy of critical mathematics education*. Kluwer Academic Publishers.
Skovsmose, O. (2005). *Travelling through education: Uncertainty, mathematics, responsibility*. Sense Publishers.
Skovsmose, O. (2014). *Foregrounds: Opaque stories about learning*. Sense Publishers.
Skovsmose, O. (2023). A performative interpretation of mathematics. In M. A. V. Bicudo, B. Czarnocha, M. Rosa, & M. Marciniak (Eds.), *Ongoing advancements in philosophy of mathematics education* (pp. 269–292). Springer.
Torfing, J. (1999). *New theories of discourse: Laclau, Mouffe and Žižek*. Wiley-Blackwell.
Wess, R., Klock, H., Siller, H.-S., & Greefrath, G. (2021). *Measuring professional competence for the teaching of mathematical modelling*. Springer.
Whitehead, A., & Russell, B. (1910, 1913a, 1913b). *Principia mathematica I–III*. Cambridge University Press.
Whorf, B. L. (1956). *Language, thought and reality*. Edited by J. B. Carrol. MIT Press.
William, P., & Chrisman, L. (Eds.). (1994). *Colonial discourser and post-colonial theory. A Reader*. Columbia University Press.
Wittgenstein, L. (1922). *Tractatus logico-philophicus*. With an introduction by Bertrand Russell. Routledge and Kegan Paul.
Wittgenstein, L. (1953). *Philosophical investigations*. Blackwell Publishers.
Wright, G. H. V. (1955). A biographical sketch. *The Philosophical Review, 64*(4), 527–545.

Chapter 14
Performatives

Abstract Mathematical performatives are found in all spheres of life. Mathematics does not simply depict reality, but rather formats the natural as well as the social reality that it is assumed to describe. Mathematics fabricates the mechanical worker as a gear to be incorporated in the clockwork-like production machinery. Mathematics contributes to the sanctification of norms, for instance with respect to health, beauty, and productivity. Mathematics is used for splitting up practices into a surface practice requiring a few competences, and a deep practice operating with complex sets of mathematical algorithms. Mathematics is formatting our future—for instance, by providing means for constructing climate models and for completing experimental forecasting. Mathematics and power are interacting phenomena, but mathematics tends to disguise this interaction. By means of mathematics it is possible to identify cases of injustice and, in this way, articulation of social justice also becomes one of the possible performatives of mathematics.

Keywords Mathematical performative · Worldview · Semantic projection · Algorithmic worker · Setting of norms · Body mass index (BMI) · Productivity · Objectivity · Neutrality · Formatting the future · Risk · Surface practice · Deep practice · Mathematics and power · Power · Social justice · Literacy · Mathemacy

In order to illustrate the significance of a performative interpretation of mathematics, I will consider different examples of mathematical performatives.

When the performative interpretation of language was introduced within the frame of analytic philosophy, particular attention was paid to expressions like "promising", "baptising", and "joking". When making a promise, one is doing more than uttering a sentence. One is putting oneself under an obligation; one is acting by promising something. Soon it was acknowledged that any kind of linguistic expression is performative. There is no performative-free language. Speaking means speech-acting. Discourse theory made a tremendous addition to the performative interpretation of language by showing how language might form worldviews and eventually our life-worlds.

The performative interpretation of language has inspired my performative interpretation of mathematics, and I want to show how mathematical performatives might

O. Skovsmose, *Critical Philosophy of Mathematics*, Advances in Mathematics Education, https://doi.org/10.1007/978-3-031-71375-0_14

be shaping, fabricating, sanctifying, splitting, forming, disguising, and articulating something.

I am not trying to indicate that some particular pieces of mathematics can be associated with particular types of performatives. In one situation a piece of mathematics might be brought into one kind of action, while in a different situation the very same mathematics might be brought into a quite different kind of action. There is no performative-free kind of mathematics, and all mathematical performatives are of a contested nature.

14.1 Shaping Worldviews

Galileo Galilei made a distinction between appearance and reality and claimed that in order to describe reality, mathematics was necessary. Like Galilei, both René Descartes and Isaac Newton found that the descriptive capacities of mathematics were due to the fact that God had created nature according to a mathematical pattern. For grasping this pattern, mathematics provided the only adequate means.

After the belief in God's existence was expelled from scientific theorising, the assumption continued the same: mathematics is the proper language for describing nature, as nature as such operates according to a mathematical pattern. Nature was seen as a mathematics-mechanical structure, and mathematics was seen as being the proper means for depicting this structure. According to a material-mechanical view, the functioning of nature can be metaphorically compared with the functioning of a clockwork mechanism where interactions occur through direct contact between components; in a clockwork there are no interactions across distances. Newton assigned gravity a particular role to play, and the functioning of gravity cannot be expressed in a material-mechanical way, but it can be expressed in mathematical terms. Therefore, I do not talk about a mechanical worldview, but about a mathematics-mechanical worldview.

In the *Tractatus Logic-Philosophicus*, Ludwig Wittgenstein (1922) formulated his picture theory of language. This theory can be condensed into an ontological and an epistemological claim. The ontological claim is that the world is composed not of things, but of facts (§1). Furthermore, Wittgenstein found that facts are organised in a logical order, which he referred to as their logical form. His epistemological claim is that the fact that the language of mathematics can picture reality is due to the fact that the language of mathematics shares its logical form with reality. Furthermore, according to Wittgenstein, the mathematical language is the only language with this property.[1] In *The Logical Structure of the World*, the original German version of which was published in 1928, Rudolf Carnap elaborated on the same ideas. Carnap (2003) presented how the construction of the world follows a logical pattern, and like Wittgenstein, he found that in order to grasp this pattern, one needs a language

[1] For a presentation of Wittgenstein's picture theory of language, see also Sect. 13.2.

constructed according to the same pattern. Carnap agreed with Wittgenstein: this language is mathematics.[2]

The basic idea shared by Galilei, Wittgenstein, and Carnap is that mathematics reflects the logical structure of reality. Consequently, mathematical descriptions have qualities that make them superior to other forms of descriptions.

However, there is a different way of looking at the mathematics-reality relationship. When mathematics is used for describing, it also shapes what it describes. When Galilei described heat, he referred to some miniscule particles. He saw these particles as having particular mathematical properties in terms of shape, size, position, and movement. By making this claim, the semantics of mathematics becomes projected onto reality. We are dealing with a *semantic projection*, where mathematical conceptions form what we assume we are merely describing.

A semantic projection is a direct expression of a mathematical performative. Mathematics and reality are not two separate entities, but rather interacting entities. Mathematics is shaping our views of the world and eventually our life-worlds.

14.2 Fabricating the Algorithmic Workplace

When we are addressing social phenomena, mathematics-based semantic projections form what we try to describe. The very social reality assumes a mathematics-mechanical format.

Mathematics has been applied in scientific management, aiming at efficiently optimising processes of production and fitting workers in accordingly. Workers are depicted as gears in the production machinery. Scientific management analyses production processes into their atomic components, and each component can be attempted optimised. The optimised components can then be assembled into the most efficient production machinery. The mechanisation of the worker becomes a part of this optimisation. The workers' movements and physical actions are studied, while his or her human features are ignored. Just as a natural phenomenon is captured in terms of shape, size, position, and movement, a worker is identified by his or her measurable physical properties. To me, scientific management illustrates how a workplace becomes configured as a mathematics-mechanical system, and how the algorithmic worker is fabricated to fit in with that system.

Such a fabrication was presented by Frederick W. Taylor in *The Principles of Scientific Management*, which first appeared in 1911. In this book, Taylor (2006) presented the worker, Schmidt, who became subjected to the principles of scientific management. The particular task that Schmidt was assigned was to load pig iron onto a railway wagon. Taylor and his staff had observed that, given the present

---

[2]Carnap (1937) gives a careful account of the nature of the logic-mathematical language that can count as the language of science. Schürmann (2023) discusses mathematical modelling with reference to Carnap's conception of the relationship between mathematics and reality.

organisation of the task, an average worker could be expected to load around 12 tons of pig iron per day, but that a different organisation of the task would make it possible for a worker to load around 47 tons. According to Taylor: "The task before us, then, narrowed itself down to getting Schmidt to handle 47 tons of pig iron per day and making him glad to do it" (p. 20).

Taylor presented the conversation he had with Schmidt, where he asked Schmidt if he wanted to become a "high-priced man", which meant a raise in salary from 1.15 US Dollars per day to 1.85 US Dollars. Unsurprisingly, Schmidt was ready, and Taylor presented the conditions:

> Well, if you are a high-priced man, you will do exactly as this man [the work leader] tells you to-morrow, from morning to night. When he tells you to pick up a pig and walk, you pick it up and walk, and when he tells you to sit does and rest, you sit down. You do that straight through the day. (p. 21)

A high-priced man does exactly what he becomes prescribed to do. He does not ask questions. He behaves like a robot. Schmidt was conceptualised as a mechanical entity, whose movements could be optimised. Schmidt turned into a cog in the clockwork process of loading pig iron.

There are only a few figures and numbers in Taylor's initial presentation of scientific management, but its principal approach becomes clear: to operate with a clear distinction between what to consider primary and what to consider secondary when optimising a work process—the same distinction applied by Galilei when distinguishing between appearance and reality. To depict the reality of a work process, one must focus on what can be captured by numbers, and disregard all secondary phenomena as human emotions and priorities. Schmidt is stripped of his human qualities; what is left is only a physical body.

In the film *Modern Times*, Charlie Chaplin presented an ironic image of the algorithmic worker. As a worker in a factory, Chaplin became transfigured into a mechanical cog in a production machine. This machine had to be optimised, which also included specification of what, when, and how the workers need to eat. In one scene from *Modern Times*, we see how a feeding machine might function. In another scene, Chaplin got literally swallowed up by the machinery.

More recently, scientific management has developed enormously by taking on a digital format. Today, many production processes are designed by computerised algorithms, and the optimisation of a production process comes to depend upon the optimisation of algorithmic complexities. This optimisation includes a development of new forms of deconstructing production processes into algorithmic unities, new ways of dividing tasks between real robots and robotised workers, and the development of faster and faster algorithms. Mathematics contributes to the formatting of the algorithmic workplace and of the workers that are needed to operate it.[3]

The very processes of leadership and management also become algorithmicised. One finds algorithmic ways of evaluating the individual workers' performance; algorithmic ways of ranking candidates for a new job; and algorithmic ways of

[3] See, for instance, Holford (2020).

estimating workers' moods and team dynamics.[4] Gig working is a new phenomenon, associated with the algorithmic workplace. A gig worker is a temporarily contractor working on an online platform. The gig worker is contracted according to the actual need of a company, which might wary from season to season, from day to day, from hour to hour.[5] At the algorithmic workplace we find the digital panopticon in full operation. Through algorithmic processes, all kinds of information about workers' performance are automatically collected. This becomes a powerful way of disciplining workers in terms of registering, rating, rewarding, or replacing.[6]

Today, not only workplaces but our life-worlds are formed through all kinds of algorithmic procedures. We are submerged in a wilderness of algorithms put into operation. While Chaplin got swallowed up by a mechanical machine, we get swallowed up by Turing machines.

14.3 Sanctifying Norms

The results of a blood test can be expressed in numbers showing the levels of sodium, cholesterol, magnesium, creatinine, potassium, urea, vitamin B-12, and many other substances. The numbers are expected to be located within specific intervals defining normality. If a number is located outside the interval, it might indicate that some treatment is necessary: some pills might be prescribed or some additional medical investigations could be required. The specification of norms draws on a range of statistical techniques, defining for instance the lower and upper bounds of the interval that defines normality.

Mathematics contributes not only to medical norm-setting, but also to specifications of norms with respect to "beauty". The Body Mass Index (BMI) can be considered a health index, but it also sets standards for the assumed ideal body. In Antiquity, visions of the beautiful body were expressed in art, such as in the sculptures of Adonis. In the famous drawing from around 1490 of the Vitruvian Man, Leonardo da Vinci presented what, at the time, was considered ideal proportions of the human body. He inscribed the body in a square and a circle, indicating for instance that the height of the person should be the same as the lengths of the person's outstretched arms (termed a person's "fathom", from the Old English "fæthm", meaning outstretched arms).

Considering such artistic expressions of the beautiful body, it appears a vulgar simplification to operate with BMI, defined as:

$$\mathrm{BMI} = w/h^2$$

[4]See, for instance, Dellot (2017).

[5]See, for instance, Jarrahi and Sutherland (2019).

[6]See, for instance, Kellogg et al. (2020).

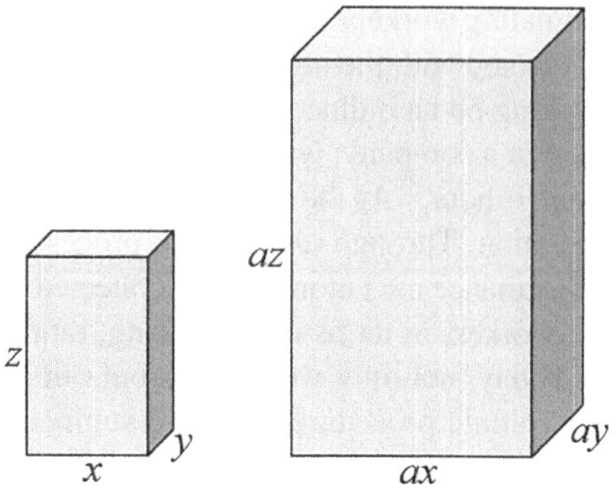

Fig. 14.1 Two proportional bodies

where w refers the person's weight measured in kilos and h to the person's height measured in metres.[7] The BMI is considered normal if it is located in the interval between 18.5 and 25. A person is called underweight if the index is smaller than 18.5, and overweight if the index is bigger than 25. Obesity is associated with an index bigger than 30.

BMI can be considered an expression of the relationship between the actual volume of a person (proportional to the weight of the person) and the idealised volume defined by the bone structure of the person. It feels consequential to assume that proportional bodies should have the same body mass indexes. However, let us assume that the bigger body is a box a times the size of the smaller box (a being a constant bigger that 1.00) as shown in Fig. 14.1.

The Body Mass Index of the smaller body would be $\text{BMI} = w/h^2$, while the index of the bigger body would be $\text{BMI} = a^3w/(ah)^2 = aw/h^2$. The body masses are not the same. If we instead consider the Body Mass Index to be defined as $\text{BMI} = w/h^3$, the body mass of the smallest body would be $\text{BMI} = w/h^3$, while the body mass of the bigger body would be $\text{BMI} = a^3w/(ah)^3 = w/h^3$. So why use h^2 and not h^3 in the definition of the BMI?

Let us consider a body being the shape of a box with height h, broadness b, and depth d, as shown in Fig. 14.2.

The Body Mass Index could then be defined as $\text{BMI} = w/hbd$. The height h of a person is straightforward to measure; the "broadness" b of a person could be considered to be broadness across the shoulders, and the "depth" of a person could be the depth of their chest (or maybe correlated with the size of their shoes). However, such a definition brings about a practical problem: while the height of a person is easy to measure, the "broadness" and "depth" are not straightforward measures. However, if we consider the broadness of a person to be proportional with the fathom of a person, and the fathom to be the same as the height of the person (as indicated by da Vinci in the drawing of the Vitruvian Man), and furthermore the depth of everybody to be the same, then we arrive at the formula $\text{BMI} = w/h^2$.

[7] See Eknoyan (2008) for an account of the historical development of the formulation of the BMI. See also Hall and Barwell (2021) and Skovsmose (2023).

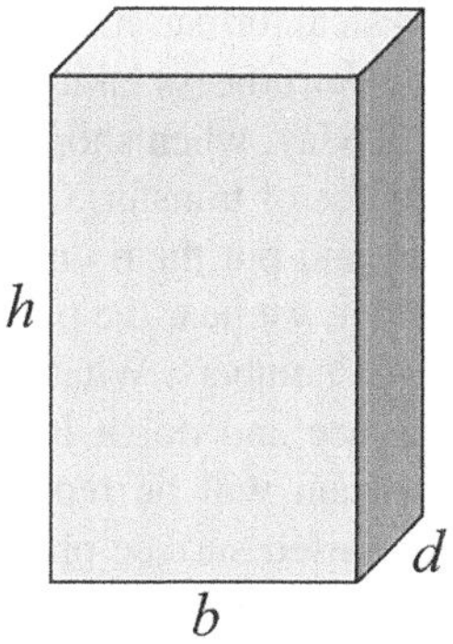

Fig. 14.2 A body in the shape of a box

The definition of the Body Mass Index as BMI $= w/h^2$ is far from being a straightforward description of a biological phenomenon. It is rather a mathematical caricature of "beauty".

This caricature is then used for indicating non-normalities in need of treatment. It set standards into which a person struggles to adjust his or her body to fit. In 1969, when Arnold Schwarzenegger won the Mr. Universe Competition, his weight was 154 kilograms and his height was 182 cm. This gives his BMI as 46. This appears to be a severe case of obesity.

Norms have also been set with respect to measures of productivity. As an example, I can refer to a measure of research productivity that was implemented for a period at Aalborg University, which I have experienced myself. Each year a researcher was ascribed a number determined by the researcher's publications over the year. A publication in, say, *For the Learning of Mathematics* would count as 1.00 point. If the publication had two authors, the first author would get 0.53 points and the second author 0.47 points. All journals were classified in three groups: some received 2.00 points, others received 1.00 point, and the rest received 0.00. The sum of the production points for the academic staff of the department had an impact on the funding the department received. Highly productive departments would be allocated some extra money compare to less productive departments. We have arrived at a mathematical caricature of "doing research".

Making mathematical caricatures is a paradigmatic component of a norm-setting mathematical performative. Norms expressed mathematically gets sanctified.

14.4 Splitting Practices

Long ago, when shopping in the supermarket, I would collect the items that I wanted to buy and bring them to the cashier, who would add up what I needed to pay. I would pass over some notes, and would receive some cash back. I had paid, and the items were mine. I had transferred money to the supermarket, and the transfer had a physical format: money in exchange for goods. One could follow with one's eyes the transfer process taking place across the counter; it was a direct physical transfer.

Today, when shopping in the supermarket, we are dealing with a quite different process of transfer. Getting to the cash point, we take the items from the shopping cart and put them on a small belt; the shop assistant passes the items through the reader, we hear the pling-sound, we insert the credit card into a reader and punch in some numbers. What we can call the *surface practice* is made up of things that we can see and do. It includes a series of physical movements, which for the shop assistant will be repeated many times during the working hours. However, this routinised surface practice is only made possible due to the installation of a huge and advanced system of mathematical algorithms encapsulated in the relevant hardware. This constitutes the *deep practice* of paying at the supermarket. Bar-codes have to be designed; they have to be read; information about the quantity of sold items has to be collected in order to keep up the stock at the supermarket; the amount to be paid has to be transferred from the account associated with the credit card to the account of the supermarket. I activate the transfer process by pressing some buttons, but I do not participate in the process. I have no chance of following the process with my eyes. It is completed by means of pre-designed mathematical algorithms. Amazon has taken this even further, where the shopper just walks in, puts items in his or her bag and walks out, by which time the money had been transferred.

We are dealing with a principal phenomenon, namely the *splitting of a practice.* In its simplified form it consists of a split between a surface practice and a deep practice. The split is made possible due to mathematics.[8]

A splitting of practice might have huge impacts, for instance with respect to efficiency. Imagine a supermarket with a long queue of consumers with full shopping carts waiting at the cashpoint, and imagine the shop assistant taking up a pencil and making a note of the prices of the items to add up the total to be paid.

The splitting of practices also means a split of competences. While advanced mathematical competences are necessary for further developing the deep practice, no such competences are relevant for improving the surface practice. Imagine that a cashier at a supermarket wants to improve his or her skills in attending to the customers and, in order to do so, starts studying the mathematical algorithms that are brought into operation at the cashpoint. This would make no sense for improving the surface practice. For the shop assistant it is important to be able to communicate in a nice way with the customers, to be able to pass the items in the proper way through the reader, and to do the same movements day after day without getting problems with his or her back and shoulders.

In *Capital*, Marx observed with respect to factory work that it "exhausts the nervous system to the uttermost; at the same time, it does away with the many-sided play of the muscles, and confiscates every atom for freedom, both in bodily and in

[8]Jablonka and Gellert (2007) elaborate on the same point in terms of mathematisation and demathematisation.

intellectual activity" (p. 548). And he adds that "the machine does not free the worker from the work, but rather deprives the work itself of all content" (p. 548).[9] Marx made these observations with factory work in mind; however, the remarks seem relevant in many cases where a split of practices results in a surface practice composed of few simple routines, and a deep practice manageable only by specialists. The formation of the algorithmic worker can be seen as a formation of a worker that fits into routinised surface practices.

The splitting of a practice is an example of a mathematical performative. By means of algorithms brought into operation, mathematics is configuring a deep practice that is doing the job, leaving the surface practice to be taken care of by low-qualified employees.

At the stock market today, clusters of algorithms are brought into operation. Algorithmic trading is an example of a mathematical performative. It includes decisions about when and what to sell and buy. We are seeing decisions and actions conducted by algorithms loaded with predefined instructions about what to do under what circumstances. When the situation is calculated to be optimal, action can be executed immediately. A study made by Robert Kissell (2020) mentions that in 2019, around 92% of the transitions at the foreign exchange market were algorithmic. We are again dealing with an example of the splitting of a practice. The selling-buying algorithms are programmed with a range of priorities. However, when a multitude of such programmes are operating in the same market, it becomes impossible to anticipate how the whole system might be functioning. One cannot do any testing in advance of the algorithmic stock market; the stock market itself becomes the site for testing.

The split between a surface practice and a deep practice also concerns the nature of what can go wrong. The risks associated with a surface practice might be quite different from those associated to a deep practice. This applies to the supermarket as well as to the stock market. In the book, *Weapons of Math Destruction*, Catherine O'Neil (2016) made a broadly accessible presentation of her interpretation of the economic collapse that took place in 2008. Her conclusion was that one factor in his collaps, maybe the most important one, was algorithmic trading. Decisions about selling and buying were executed with such a speed that the whole market got out of control.

Automatic pilot is another example of a split of practices. During a long-distance flight, the automatic pilot takes over, while the real pilot is in charge during take-off and landing. The real pilot is on board in this case, but there are also cases like certain military operations where one operates drones with nobody on board. This possibility is likewise being implemented for civil purposes in small aeroplanes used for the transport of goods. Money is saved on fuel by reducing the total weight of the aeroplane by the weight of the pilot. In the future, one can imagine passenger aeroplanes without any pilot on board, just as we now find metros with no driver

[9]The observations made by Marx have already been referred to in Sect. 3.2.

on board.[10] Making this possibility feasible is one of the hot topics in aeronautic research today. In the case of automatic piloting, we are dealing with a split of expertise practices. The deep practice is configured through a most advanced configuration of algorithms. The corresponding surface practice, however, is not a simple practice at all. It is also drawing on advanced mathematics brought into action.

Whatever split of practice we are dealing with, it is an example of mathematical performatives. It reshapes everyday practices and in doing so, reshapes our life-worlds.

14.5 Forming Futures

Mathematics is forming our conception of the physical world by moulding it as a mathematics-mechanical system. Mathematics is also forming social phenomena according to mathematics-mechanical patterns. But in what way might mathematics be forming our future?

Making a prediction of the future is a common desire. Reading a horoscope was one way of trying to do so, as it was assumed that celestial constellations had an impact on what took place on earth. Weather forecasting has been a daily practice during the history of humanity. Successful farming depends on weather forecasting, and fishermen have always learned to read the sky. Today, weather forecasting has become an advanced mathematical discipline.

As we consider the future weather of the planet, we come to address possible climate change, and to do such forecasting, mathematics is indispensable. There is no commonsense parallel to mathematics-based predictions of climate change. A climate model is a construction including a range of parameters that are assumed to be significant. Such parameters could concern the dynamics of the atmosphere, the oceans, and the ice sheets. The dynamics of the ocean might, for instance, depend on the degree of salinity, as well as the dynamics of the ice. The complex system of interacting parameters is connected by equations, representing the connections that one assumes to be operating.[11]

The mathematical format of a climate model makes it possible to conduct *experimental forecasting*. Such forecasting starts by ascribing a set of values to the parameters of the model and then calculating what the output of the model will be. Then one changes the value of one or more of the parameters and sees what the new output might be. In this way one gets an idea of what parameters have a significant impact on the output of the model. An experimental forecast might indicate the parameters on which the climate of the future appears to depends the most.

[10] See, for instance, Pilotless Passenger Planes—How Far Away Are They? (simpleflying.com).

[11] For discussions of climate models, see, for instance, MacKenzie (2007).

Experimental forecasting with respect to climate change is a contested activity. The use of mathematics does not nominate a particular climate model as being the "correct" one. A mathematical climate model is not "similar" to any physical, biological, or environmental mechanism; it is rather a conglomerate of postulated relationships assumed to demonstrate some similarity with real-life processes. What parameter is selected and what calculation machinery is created might reflect particular political, industrial, or environmental interests. Mathematics does not ensure any form of objectivity or neutrality with respect to experimental forecasting, but rather adds to its contested nature. Still, mathematics is a necessary means for carrying out experimental forecasting.

Due to experimental forecasting, mathematics formats conceptions of the future. Environmental problems can be portrayed as possible to deal with in a rational and technological way. Some parameters might be pointed out as being the main ones "responsible" for climate change, and a solution to future climate problems could be described as keeping the value of these parameters within certain limits. This might be helpful, but to assume that we are dealing with a solution might be an illusion produced by the very mathematics-based experimental forecasting itself.

Mathematics-based experimental forecasting not only formats our conceptions of the future, it also formats the future itself. The configuration of risks that condition our future are complex constructions to which mathematics-based experimental forecasting is contributing. One can conceptualise risk as a product $R(A) = CP$, where $R(A)$ is the risk associate to an event A taking place, C is the possible consequence of A taking place, and P is the probability that A in fact takes place. A risk might be considered small, even if the possible consequences are enormous. The *Rasmussen Report* was first published in 1975. The aim was to provide a numerical estimation of the likelihood that a disaster might take place at an atomic power plant.[12] The estimation could not be based on any experimental observations, it must be based totally on mathematical modelling. The occurrence of an accident could be depicted as a sequence of events, all of which were going wrong: an instrument did not function, another instrument was out of order; a technician overlooked something: another made a wrong decision; and so on. It was assumed that if one were able to identify all possible sequences of events that would lead to a disaster, one could provide an estimation of the likelihood that an accident occurs at a nuclear power plant. The *Rasmussen Report* concluded that the likelihood of an accident taking place was extremely small. The report achieved much credibility, as it apparently provided a solid mathematics-based estimation of a risk factor. Consequently, it had a tremendous impact on the future by contributing to the choices of energy policy in the USA, as well as in many other countries.

The partial meltdowns that happened at the Three Mile Island power plant in the USA in 1979, the Chernobyl disaster that took place in 1986 in the Soviet Union

[12]The title of the report is *Reactor Safety Study: An Assessment of the Accident Risk in the U.S. Commercial Nuclear Power Plants*. An *Executive Summary* is available at https://www.osti.gov/servlets/purl/7134131

(now Ukraine), and the disaster that happened in Fukushima in Japan in 2011 all indicate that the conclusions of the *Rasmussen Report* are questionable despite their presumed credibility.

Risks can be produced today, without having any visible immediate impact. Risks only represent propensities, but these propensities are forming part of the configuration of the future. One way of conceptualising, addressing, or ignoring risks is through mathematics-based experimental forecasting. In this way mathematics formats conceptions of the future. It forms decisions and actions concerning the future. In the end, it forms our future.

14.6 Disguising Powers

Through mathematics-based actions, mathematics and power become integrated. Mathematics tends to disguise this integration. Power structures become camouflaged behind the appearance of mathematics as being objective and neutral. By objectivity I refer to the idea that, by means of mathematics, one can depict reality as it really is. That mathematics is neutral refers to the idea that no value judgements are articulated through mathematics. In being objective and neutral, no particular interests seem to be exercised through mathematics.

The claim that mathematic is objective and neutral is compressed into the ideology of certainty.[13] This ideology can be summarised by two set of ideas: first, mathematics represents absolute truth, which does not depend on any empirical observations; second, mathematics can be applied to all kinds of real-life problems, and the mathematics-based solutions to such problems can be considered objective and neutral. The ideology of certainty assigns a high degree of reliability to statements expressed mathematically. Thus, it removes the discussion of the mathematics-power relationships from the agenda of mathematical practices—whether in research, application, or education—as mathematics-soaked statements become seen as expressions of truth rather than of power.

The ideology of certainty makes it acceptable, if not consequential, not to address the social complexities within which mathematics is involved. According to an ideology of certainty there is nothing controversial to be associated with respect to such involvements. To a critical philosophy of mathematics, it is important to challenge the ideology of certainty and to reveal the many ways that power and mathematics are interacting.

[13] See Borba and Skovsmose (1997).

14.7 Articulating Social Justice

Literacy refers to the capacity to read and write. However, Paulo Freire gave the notion a much broader interpretation; it also came to include the capacity for grasping social and political complexities and to act against political oppression and economic exploitation. Literacy not only concerns descriptions of states of affairs, but also the altering of states of affairs. Understood in this way, literacy is a direct expression of a performative interpretation of language.

Inspired by Freire's interpretation of literacy, in *Towards a Philosophy of Critical Mathematics Education* I suggested a similar interpretation of mathemacy.[14] Mathemacy refers to a competence in grasping mathematical notions and ideas and a mastery of mathematical techniques, including the ability to carry out mathematical calculations. Furthermore, mathemacy refers to a capability in applying mathematics to real-life problems; it refers to an ability in addressing social states of affairs, identifying cases of social injustice, and making socio-political changes. The notion of mathemacy embraces a performative interpretation of mathematics.

The performatives encompassed in the notion of mathemacy have been expressed and further elaborated by the concept of "reading and writing the world with mathematics". This expression was used by Eric Gutstein (2006) with direct reference to Freire. By reading the world with mathematics, Gutstein refers to the capability of using mathematics to interpret what is considered to be a social "given" as, in fact, being open to change. By reading the world with mathematics one might become aware of patterns of social exclusion, of degrees of economic inequality, of the intensity of economic exploitation of workers, of the extension of sexism, and of the impact of racism. By writing the world with mathematics, Gutstein refers to the capacity to use mathematics to try to implement social and political changes. Mathematics can be brought into action to change how we see reality, as well as to change reality itself. The expression of reading and writing the world with mathematics is a direct expression of a performative interpretation of mathematics.

By reading and writing the world with mathematics, one can work for social justice. This is the ambition of Gutstein's educational project. The expression of reading and writing the world with mathematics is coined within the outlook of critical mathematics education. However, interpreting and changing the world by means of mathematics can also be applied within quite different outlooks. What can be obtained by means of mathematics within scientific management is an example. Making a production process more efficient can hardly be seen as part of a struggle for social justice, but rather as a concern for accumulating profit. Still, scientific management is based on a mathematical reading of an existing work process, and on mathematics-based identification of possible alternatives.

[14] See Skovsmose (1994). See Chronaki (2010) for further elaborations of the notion of mathemacy.

Mathematical performatives are of contested nature. The same piece of mathematics can be brought into action to serve very different interests. This is important to have in mind when presenting an articulation of social justice as a possible mathematical performative. Mathematics might as well be articulating opportunities for exploitation and oppression. Articulating social justice is just one possibility among many dissimilar mathematical performatives.

References

Borba, M., & Skovsmose, O. (1997). The ideology of certainty. *For the Learning of Mathematics, 17*(3), 17–23.

Carnap, R. (1937). *The logical syntax of language*. Routledge and Kegan Paul.

Carnap, R. (2003). *The logical structure of the world*. Open Court.

Chronaki, A. (2010). Revisiting mathemacy: A process-reading of critical mathematics education. In H. Alrø, O. Ravn, & P. Valero (Eds.), *Critical mathematics education: Past, present and future* (pp. 31–49). Sense.

Dellot, B. (2017). The algorithmic workplace need not be a dystopia. *RSA Blog*. https://www.thersa.org/blog/2017/11/the-algorithmic-workplace-need-not-be-a-dystopia

Eknoyan, G. (2008). Adolphe Quetelet (1796–1974): The average man and indices of obesity. *Nephrology Dialysis Transplantation, 23*(1), 47–51.

Gutstein, E. (2006). *Reading and writing the world with mathematics: Toward a pedagogy for social justice*. Routledge.

Hall, J., & Barwell, R. (2021). The mathematical formatting of obesity in public health discourse. In A. Andersson & R. Barwell (Eds.), *Applying critical mathematics education* (pp. 210–228). Brill and Sense Publishers.

Holford, D. (2020). The algorithmic workplace and its enactive effect on the future of professions. *Futures, 122*, 102609.

Jablonka, E., & Gellert, U. (2007). Mathematisation – demathematisation. In U. Gellert & E. Jablonka (Eds.), *Mathematization and de-mathematization: Social, philosophical and educational ramifications* (pp. 1–18). Sense Publishers.

Jarrahi, M. H., & Sutherland, W. (2019). Algorithmic management and algorithmic competencies: Understanding and appropriating algorithms in gig work. In N. Taylor, C. Christian-Lamb, M. Martin, & B. Nardi (Eds.), *Information in contemporary society conference 2019* (Lecture Notes in Computer Science, vol 11420). Springer. https://doi.org/10.1007/978-3-030-15742-5_55

Kellogg, K. C., Valentine, M., & Christin, A. (2020). Algorithms at work: The new contested terrain of control. *The Academy of Management Annals, 14*(1), 366–410. https://doi.org/10.5465/annals.2018.0174

Kissell, R. (2020). *Algorithmic trading methods: Applications using advanced statistics, optimization, and machine learning techniques*. Academic.

MacKenzie, D. (2007). *Mathematics of climate change: A new discipline for an uncertain century*. Mathematical Sciences Research Institute. http://library.msri.org/msri/MathClimate.pdf.

O'Neil, C. (2016). *Weapons of math destruction: How big date increase inequality and threatens democracy*. Broadway Books.

Schürmann, U. (2023). Mathematical modelling: A philosophy of science perspective. In M. A. Bicudo, C. Bronislaw, M. Rosa, & M. Marciniak (Eds.), *Ongoing advancements in philosophy of mathematics education* (pp. 309–329). Springer.

Skovsmose, O. (1994). *Towards a philosophy of critical mathematics education*. Kluwer Academic Publishers.

Skovsmose, O. (2023). *Critical mathematics education*. Springer.

Taylor, F. W. (2006). *The principles of scientific management*. Cosimo Classics.

Wittgenstein, L. (1922). *Tractatus logico-philophicus*. With an introduction by Bertrand Russell. Routledge and Kegan Paul.

Chapter 15
Action and Ethics

Abstract Much philosophy of action has focused on what could be called individual or collective actions. One can, however, also consider structural actions, which need not be deliberately performed by any individual, team, group, or institution. Structural actions can be extremely powerful and have a profound and also devastating social impact. In this chapter I am going to elaborate on the performative interpretation of mathematics by relating mathematics to individual, collective, as well as structural actions. A variety of different consequences can be brought about by actions, and so these actions need to be critically addressed. I see critique of action as being an ethical endeavour. Logical positivism has insisted on the existence of a sharp distinction between, on the one hand, mathematics and science, and, on the other hand, ethics. Logical positivism inserts mathematics and science into an ethical vacuum. A critical philosophy of mathematics opposes the elimination of ethics from mathematics, and highlights that mathematics is involved in all spheres of life, including the formation of wonders as well as of horrors. This brings mathematics face to face with an ethical challenge.

Keywords Performative interpretation of mathematics · Structural racism · Structural action · Mathematics and power · Ethics · Logical positivism · Metaphysics · Boo-hoorah theory

The performative interpretation of mathematics makes it important to more carefully consider the notion of action, and in particular the relationships between action, ethics, and responsibility. This forces us to recognise an ethical challenge that concerns all mathematical practice, whether we are dealing with research, application, or education.

The philosophy of mathematics needs to address ethical issues, as is discussed in the book *Connecting Humans to Equations: A Reinterpretation of the Philosophy of Mathematics*, where Ole Ravn and I introduce an ethical dimension in the philosophy of mathematics in addition to an ontological, epistemological, and sociological dimension (Ravn and Skovsmose, 2019). This book is a reworking and translation of the book *Matematikfilosofi* from 2011 written in Danish (Skovsmose and Ravn, 2011). I think of *Matematikfilosofi* as the initiation of the ethical turn in the

O. Skovsmose, *Critical Philosophy of Mathematics*, Advances in Mathematics Education, https://doi.org/10.1007/978-3-031-71375-0_15

philosophy of mathematics. Paul Ernest has also contributed to this turn, not only through studies in the philosophy of mathematics, but also in the philosophy of mathematics education (see Ernest, 2016, 2018, 2021, 2024).

To recognise that mathematics is formed by indefinite constructions, combined with the performative interpretation, is pivotal for the ethical turn in the philosophy of mathematics. This turn is essential to a critical philosophy of mathematics.

15.1 Action

The philosophy of action is a huge domain, which includes many attempts to define the notion of action.[1] An "action" has been distinguished from a "happening". I can decide to take a walk; this is my decision; and the walking will be my action. When walking I might stumble over a stone. This is something that happens to me; this is not what I had in mind; this is not my action. What, then, is the difference between an action and a happening?

One way to expose this difference is to pay attention to the notion of *intentionality*. One could claim that an action is characterised by the person intending something; an action is directed, an action has a purpose. I had the intention of taking a walk, but no intention of stumbling over a stone.[2] In philosophy one finds differentiations between intentionality and intention, but I am going to leave aside any such differentiation and use the notions interchangeably. (Naturally one can claim that characterising actions by intentionalities is merely a transposition of the problem of definition. Instead of having to clarify the notion of action, one needs to clarify the notion of intentionality, which appears to be just as vague as that of action.)

Much philosophy of action has focused on what could be called *individual* actions. They are actions performed by an acting subject. Characterising individual actions by means of the notion of intentionality seems to work quite well. A person can have intentions that direct his or her actions.

One can also consider *collective* actions, which can be exemplified by any kind of team work, such as the achievements of a football team. A collective action can be carried out by a team, an institution, a political party, or a government. We also know that a collective, such as a government, can stumble over an unexpected event, such as the unfortunate behaviour of one of its members. Consequently, one might also need to distinguish between an action and a happening with respect to collective activities. As with individual actions, so can collective actions be associated to an acting subject, though in this case the subject would be a collective rather than an individual.

[1] See O'Connor and Sandis (2010) and Paul (2020).

[2] For a discussion of the notion of intentionality, see Skovsmose (2014).

One can consider the relevance of talking about *structural actions*, without any well-defined acting subject. My inspiration for introducing this term comes from the concept of structural racism.[3] Through this notion it is highlighted that racism is not only rooted in particular racist acts, but also in established social structures and practices. As an illustration, let me refer to an example that I have already presented before.[4] I have been hospitalised several times here in São Paulo. I have never experienced any case of explicit racism, where a member of staff talks negatively about others due to the fact that they are black (*preto* or *pardo*). Never have I witnessed any racist manifestations. However, if we look at the staff of the hospital, we will see clear differences. I have never been treated by a black doctor. All doctors that I have seen, male and female, have been white. Some of the nurses at the hospital, male and female, are *pardo*. Maybe I have seen a *preto* nurse, but I cannot clearly remember. The majority of nurses are white. A huge group of people do the cleaning at the hospital. Here the majority are *preto* or *pardo* women. This clear difference in the representation of *preto*, *pardo*, and white people in the staff of the hospital is not decided upon by some racist managers. The difference is produced by deep socio-economic structures, formed through layers of social and economic inclusion and exclusion. This differentiation is produced through the history of colonisation, slave trade, and exploitation of certain groups of people. Racist differentiation is formed by the unequal distribution of economic resources, including unequal access to education. We are dealing with an example of structural racism.

Having the notion of structural racism in mind, I want to talk about structural actions that need not be deliberately performed by any single or group of individuals. Structural actions do not have a well-specified acting subject, but operate through social structures. Structural actions can be extremely powerful, and have a profound and also devastating social impact. Such actions might be rooted in individual and collective actions, but they are more than the sum of such actions. While the notion of intentionality seems to function adequately when characterising individual and collective actions, it seems uncertain what kind of intentionality might be acted out through structural actions.

I am going to elaborate on a performative interpretation of mathematics by relating mathematics and action, and I will have in mind individual, collective, as well as structural actions.

15.2 Responsibility

The close connection between action and responsibility has been pointed out by existentialist philosophy. Jean-Paul Sartre (1989) described human beings as being free to make their own choices. Human beings are not things. They are nothing. They

[3] See, for instance, Almeida (2019).

[4] See Skovsmose (2023).

are nothing except what they act out and become. Human beings have no essence. This existentialist idea has been condensed into the dictum that existence comes before essence. We form ourselves by choosing what we do. Our existence comes before our self-constructed essence. We are "doomed to be free", and consequently we are responsible for our actions.

I am not trying to embark on existentialist philosophy here, and I do not operate with the existentialist conception of freedom. However, I fully recognise the importance of exploring the relationship between action and responsibility. This relationship appears quite obvious when we are dealing with individual and collective actions. In both cases the acting subject, being a person or a collective, can be identified and held responsible for what has been done. The situation seems less transparent when we are dealing with structural actions. It is difficult, if not impossible, to identify a person or a collective as being the acting subject of a structural action. Structural actions do not seem to have an acting subject. They need not be intended in any straightforward interpretation of intention. The responsibility that might be associated with structure actions seems to be floating around. Who is the performer of a structural action? Who can be held responsible?

The issue of responsibility with respect to structural actions is crucial when we are addressing mathematical performatives. Such performatives can be individual, collective, but also structural. The clarification of what responsibility might mean with respect to structural actions becomes important when addressing possible responsibilities for the output of mathematics-based actions. This output need not be intended, but can be formed through a complex set of political, technological, sociological, as well as mathematical factors. It seems far-fetched to hold Alan Turing personally responsible for the creation of the algorithmic workplace, even though in a profound way he contributed to the understanding of an algorithm. It seems as well far-fetched to hold Godfrey Hardy responsible for the application of number theory to military cryptography. It might also appear inappropriate to hold Albert Einstein personally responsible for the creation of the atomic bomb, even though he formulated the mathematical formula $E = mc^2$ that compressed the insight that made the bomb conceivable. The relationship between mathematics-based structural actions and responsibility is blurred.

Mathematical war research could be relevant for antagonist military powers. We are faced with a paradigmatic example of mathematical applications being indefinite. However, the concept of responsibility needs to be handled with great care, precisely because applications of mathematics are indefinite. In classifying research as mathematical war research, it is irrelevant whether the mathematicians themselves are aware of the military significance of what they are doing. The mathematical community cannot escape responsibility by maintaining a strict focus on intrinsic mathematical problems. Responsibility cannot be eliminated by the mathematical community ignoring what they might be involved in.

I see responsibility in terms of a readiness to take on the possible consequences of bringing mathematics into action in the public agenda. Some structural actions might be acceptable; some might be cherished; some might be repulsive; some might be horrifying. We become involved in such actions by ignoring them, by accepting them, by endorsing them. We could also question them or confront them. Maybe one cannot be responsible for a structural action as an acting subject, but one can be responsible for whether a structural action is ignored, endorsed, or confronted. This understanding of responsibility is important to bear in mind when facing the ethical challenge provoked by the performative interpretation of mathematics.

Responsibilities regarding mathematical performatives concern not only the mathematical research practice, but any practice drawing on mathematical resources. It concerns any education practice, where students not only learn mathematics, but also learn about mathematics and develop an attitude with respect to mathematical performatives.

15.3 Ethics

Whatever is brought about by an action—individual, collective, or structural—could potentially be different. The output could be most welcome, but could equally be devastating or disastrous. Since actions have the potential to be different, they need to be critically addressed, and this observation indicates the intimate connection between action and ethics.

Ethics has been parcelled up by notions such as bioethics, machine ethics, military ethics, business ethics, medical ethics, descriptive ethics, evolutionary ethics, post-Modern ethics, rational ethics, role ethics, pragmatic ethics, and virtue ethics. However, I am not interested in following up any such differentiation, but want to restrict myself to considering two overall and fundamental ethical positions: the teleological and the deontological.

According to a teleological position, an action needs to be judged according to its consequences. This idea has been elaborated upon in detail by utilitarianism, judging an action by means of pleasure and pain analyses. The deontological position confronts the teleological by highlighting that there also exist some a priori principles according to which an action needs to be judged. Could one offer up the life of a human being for the testing e new maybe dangerous medical treatment? Following a teleological position, one needs to consider how many lives one could save by offering one life. Following a deontological position, the principle that one does not offer up any human lives in order to complete medical experiments is valid quite independent of the benefits to one from doing so. The deontological perspective has deep religious roots; it has been suggested that the *Bible* provides basic guidelines for what Christianity considers acceptable behaviour. The teleological perspective, as elaborated by utilitarianism, attempts to separate ethics from religious considerations by establishing ethical questions as strict empirical questions. The deontological position can also be disconnected from religious beliefs by trying to ground overall ethical principles in some universal human rights.

I see ethical positions as social constructions that might include teleological as well as deontological lines of argumentation.[5] I can try to illustrate what this means by considering the notion of justice, which is pivotal in many ethical discussions. In the dialogue *The Republic*, Socrates discusses what justice could mean (Plato, internet). The dialogue evolves around the general assumption that the meaning of a notion like justice exists in an idealised form and that the aim of a conceptual clarification is to discover what this meaning is. Contrary to any such version of Platonism, I consider the meaning of a concept such as justice, like any other ethical notion, to be constructed. We are dealing here with a contested construction, as the meaning of justice could be constructed in different and also in contradictory ways.

By means of a thought experiment, John Rawls illustrated what it could mean to construct a conception of justice. Such a construction could take place in an assembly if two conditions were fulfilled. First, that everybody from the assembly would live in the society to which the conception of justice is going to be applied. Second, that nobody from the assembly would know in advance what position they were going to obtain: ill or healthy, man or woman, black or white, rich or poor, etc. By setting such conditions Rawls formulated idealised and impossible conditions for constructing a conception of justice.[6]

I am inspired by Rawls' thought experiment, but I want to remove it from the idealised and impossible frame and bring the construction of ethical conceptions into real-life contexts. I find it important to establish fora where ethical issues and controversies can be addressed. The ethical turn that I have in mind reaches much further out than to the philosophy of mathematics. It concerns all mathematical practice. It concerns the agenda set by mathematical research, whether we are dealing with pure or applied mathematics. It concerns mathematics education, whether we are dealing with school or university education. The classroom is also a possible forum for the construction of conceptions of social justice.

15.4 Ethics as Nonsense

Making the ethical turn in the philosophy of mathematics and in mathematical practice is a tremendous task. We are confronting a paradigmatic shift in the conception of the relationship between science including mathematics and ethics. A basic assumption of the modern conception of science is that ethics is of a completely different nature than science, and that scientific activities need to be kept strictly separate from ethical considerations.

Over time, the separation between science and ethics has found many different expressions. In *A Treatise of Human Nature*, first published in 1739–1740, David

---

[5]I have advocated for ethical constructivism: seeing ethical conceptions, including conceptions of social justice, as an expression of social construction. See Skovsmose (2012, 2018).

[6]For Rawls' theory of justice, see Rawls (1999).

Hume (1711–1776) presented a radical version of empiricist philosophy (Hume, 2009). The book, however, was largely ignored, and in 1748 Hume published *An Enquiry Concerning Human Understanding* where he recapitulated and clarified some of his previous formulations. Hume (1963) concluded the *Enquiry* in the following way:

> If we take in our hand any volume; of divinity or school metaphysics, for instance; let us ask. *Does it contain any abstract reasoning concerning quantity or number?* No. *Does it contain any experimental reasoning concerning matter of fact and existence?* No. Commit it then to the flames; for it can contain nothing but sophistry and illusion. (p. 165, italics in the original)

In this provocative way, Hume distinguished between, on the one hand, mathematical statements and statements that have an empirical content and, on the other hand, statements without any significance. Moralising philosophy and religious texts such as the Bible includes "nothing but sophistry and illusion".

In 1933, the young philosophy PhD student Alfred Ayer (1910–1989) was on his way to Vienna. His supervisor, Gilbert Ryle from Oxford University had recommended the trip. Ayer stayed in Vienna for half a year, where he attended meetings in the Vienna Circle. Back in Oxford he wrote a powerful and clear presentation of the outlook of the Vienna Circle. Ayer's book *Language, Truth, and Logic* was published in 1936; it became the principal introduction to the English-speaking world of the outlook of the Vienna Circle, which soon became known as logical positivism.

In *Language, Truth, and Logic*, Ayer (1970) sharpened Hume's formulation of ethics, endorsed by logical positivism, by presenting what became known as the Boo-Hoorah Theory. According to this theory, ethical statements are nothing but expressions of feelings of the same type as saying "Ufa" when experiencing a bad smell, and saying "Ahh" when tasting an apple strudel. Ethics has no epistemic content; it has nothing to do with science.

In 1929 the Vienna Circle had published a manifesto with the title *Wissenschaftliche Weltauffassung: Der Wiener Kreis* (*The Scientific World View: The Vienna Circle*), which condensed the outlook of the Circle (Hahn et al., internet). The Vienna Circle made a huge point of distinguishing between nonsense and what counts as meaningful scientific statements. Statements including notions like "the absolute", "things in itself", "essence", and "God" are not to be thought of as expressions of some deep insights. Instead, they have to be confronted by the question: Is it possible to identify an empirical procedure for addressing such statements? If not, they can be considered to be devoid of meaning.[7] This principle of verification leads precisely to the distinction between, one the one hand, empirical and mathematical statements and, on the other hand, meaningless statements including metaphysics and ethics.

The Scientific World View highlights that the scientific worldview only considers empirical statements and analytic statements of logic and mathematics as

[7] See Carnap (1959), the original German version of which was published in 1932.

meaningful. This conviction is engraved in the very expression "logical positivism". The scientific worldview claims that knowledge needs to be based on positive observations and that logic, including mathematics, provides the analytical approach to be used in science. The scientific worldview presents science, including mathematics, as being *detached* from political, cultural, religious, ideological, and ethical issues. I see logical positivism as launching the most ambitious project of inserting mathematics and science into an ethical vacuum.

The detachment of mathematics and science from social issues has become part of the broadly accepted working philosophies of mathematics and science. A working philosophy need not be articulated explicitly, but can be enacted through the daily practices of research, application, and education. While the position of logical positivism was that mathematics and science *ought to be* detached from ethics, today's working philosophies embrace the idea that mathematics and science *in fact* are detached and consequently will be objective and neutral. In this way working philosophies have become ideological justifications of the self-inflicted isolation of science and mathematics from political discourses.

15.5 Mathematics and the Ethical Challenge

The doctrine that mathematics is objective and neutral functions as legitimisation for mathematical research being oblivious to the possible socio-political impacts of doing mathematics. The doctrine makes it possible to dismiss ethical considerations as being irrelevant to mathematical practices. The ethical vacuum that logical positivism formed, isolated mathematics from any form of social affairs and released mathematicians from any social responsibility.

The separation between mathematics and ethics has been assumed by logicism, formalism, and intuitionism. Much subsequent philosophy of mathematics continues to operate in the same ethical vacuum as the Three Magi. As illustration, *The Oxford Handbook of Philosophy of Mathematics and Logic*, edited by Stewart Shapiro (2005), is devoid of ethical reflections. The philosophy of mathematical practice also locates itself in an ethical vacuum. What we see here is a profound paradigmatic positioning that the ethical turn in the philosophy of mathematics tries to overcome.

The ethical turn seems to imply that mathematicians must assume responsibility for what they are doing. This could make good sense in cases where mathematical performatives take the form of individual or collective actions, where a mathematician or a group of mathematicians can be identified as being the acting subject. However, mathematical performatives might also take the form of structural actions, and in this case the very notion of responsibility ends up floating.

There are different ways of trying to escape responsibility. One possibility is to claim that what one is doing is without any impact. Such an escape has been attempted by Godfrey Hardy in *A Mathematician's Apology*, first published in 1940. Hardy (1967) was concerned about the possible roles of science and mathematics in war research. He found the use of mathematics in ballistic research to be repulsive. However, with respect to what he referred to as "real mathematics" he found the situation to be quite different. A paradigmatic example of real mathematics was number theory, to which Hardy himself had contributed profoundly. Hardy claimed that number theory, like any form of real mathematics, has no utilitarian value. Consequently, real mathematicians can have a clear conscience. What they are doing cannot be applied anywhere, nor in war research. When the utilitarian value is reduced to zero, there is nothing to be responsible for. More recently, the development of modern cryptography has, however, demonstrated Hardy to be completely wrong; real mathematics might have a tremendous utilitarian value, also in war research.

Another way of trying to escape responsibility is to claim that one could not have done any differently. In 1961, during the trial that took place in Jerusalem against Adolf Eichmann, who had administrated the transportation of people to the extermination camps, he tried to escape responsibility by stating that he was just following military orders. As in this way he was being forced to do what he did, he could not be held personally responsible. This is a general pattern for escaping responsibility, and in fact "being forced to do something" is expressed in many different ways. One can feel forced to do something as one needs to keep a job and earn a salary to pay the mortgage and feed one's children; one can feel forced to do what one is doing, as this forms part of an overall project that one had been hired to participate in; and one can feel forced to do something due to pressure from colleagues, companies, and institutions.

But what about mathematical performatives taking the form of structural actions? Even though the mathematical community cannot assume responsibility for the execution of mathematics-based structural actions, it can assume responsibility for bringing, or not bringing, the possible social impact of such actions into the public agenda. The mathematical community can be held responsible for ignoring, endorsing, or confronting the possible impact of mathematics-based structural actions.

In her report on the trial of Adolf Eichmann, Hannah Arendt (1977) related that he did not appear, as one might have expected, as an ardent Nazi but as an anonymous administrator. This brought her to talk about the banality of evil. As Zygmunt Bauman (1989) observed in *Modernity and the Holocaust*: cruelty need not be committed by cruel individuals, but can be committed by ordinary men and women doing their ordinary duties.[8] Having this observation in mind, I have talked about the *banality of mathematical expertise*, referring to the all-absorbing fixation on developing mathematics while ignoring what the impact of bringing mathematics into action might be.[9] The banality of mathematical expertise manifests itself in research publications fully concentrated on presenting mathematical results, while ignoring socio-political issues related to these results; it manifests itself in the curriculum of university education when only specified in terms of mathematical topics, without preparing students to engage in reflections about the horrors and

[8] See Sect. 1.3.

[9] See Skovsmose (2020), which has been reworked as Chapter 7 in Skovsmose (2023).

wonders associated to mathematical performatives; and it manifests itself in public discourses through the absence of mathematicians' voices.

The banality of mathematical expertise is a direct expression of the ethical vacuum within which mathematics has been submerged. Filling this vacuum is urgent for a critical philosophy of mathematics and for a critical mathematics education. One place to start is in the mathematics classroom.

References

Almeida, S. (2019). *Racismo estrutural*. Editora Jandaíra.

Arendt, H. (1977). *Eichmann in Jerusalem: A report on the banality of evil*. Penguin Books.

Ayer, A. J. (1970). *Language, truth and logic*. Victor Gollancz.

Bauman, Z. (1989). *Modernity and the holocaust*. Polity Press/Blackwell Publishing.

Carnap, R. (1959). The elimination of metaphysics through logical analysis of language. In A. J. Ayer (Ed.), *Logical positivism* (pp. 60–81). The Free Press.

Ernest, P. (2016). A dialogue on the ethics of mathematics. *The Mathematical Intelligencer, 38*(3), 69–77.

Ernest, P. (2018). The ethics of mathematics: Is mathematics harmful? In P. Ernest (Ed.), *The philosophy of mathematics education today* (pp. 187–216). Springer.

Ernest, P. (2021). The ethics of mathematical practice. In B. Sriraman (Ed.), *Handbook of the history and philosophy of mathematical practice* (pp. 1–38). Springer. https://doi.org/10.1007/978-3-030-19071-2_9-1

Ernest, P. (2024). *Ethics and mathematics education: The good, the bad and the ugly*. Springer.

Hahn, H., Neurath, O., & Carnap, R. (internet). *The scientific world view: The Vienna Circle.* https://the-scientific-conception-of-the-world-the-vienna-circle (rreece.github.io)

Hardy, G. H. (1967). *A mathematician's apology*. With a foreword by C. P. Snow. Cambridge University Press.

Hume, D. (1963). *Enquiries concerning the understanding and concerning the principles of morals*. Clarendon Press.

Hume, D. (2009). *A treatise of human nature*. The Floating Press.

O'Connor, T., & Sandis, C. (2010). *A companion to the philosophy of action*. Wiley-Blackwell.

Paul, S. (2020). *Philosophy of action: A contemporary introduction*. Routledge.

Plato. *The republic*. The Internet Classics Archive. Retrieved from http://classics.mit.edu/Plato/republic.7.vi.html.

Ravn, O., & Skovsmose, O. (2019). *Connecting humans to equations: A reinterpretation of the philosophy of mathematics*. Springer.

Rawls, J. (1999). *A theory of justice*. Oxford University Press.

Sartre, J.-P. (1989). *Being and nothingness*. Translated by Hazel E. Barnes; introduction by Mary Warnock. Routledge.

Shapiro, S. (Ed.). (2005). *The Oxford handbook of philosophy of mathematics and logic*. Oxford University Press.

Skovsmose, O. (2012). Justice, foregrounds and possibilities. In T. Cotton (Ed.), *Towards an education for social justice: Ethics applied to education* (pp. 41–62). Peter Lang.

Skovsmose, O. (2014). *Foregrounds: Opaque stories about learning*. Sense Publishers.

Skovsmose, O. (2018). Critical constructivism: Interpreting mathematics education for social justice. *For the Learning of Mathematics, 38*(1), 38–44.

Skovsmose, O. (2023). *Critical mathematics education*. Springer.

Skovsmose, O., & Ravn, O. (2011). *Matematikfilosofi*. Systime.

Chapter 16
Reflective Inquiries in the Classroom

Abstract In this final chapter I present a dialogic theory of learning mathematics in terms of the following learning interacts: getting in contact, exploring, positioning, foregrounding, externalising, and doubting. This theory captures both the learning *of* mathematics and learning *about* mathematics. When having the latter kind of learning in mind, I also talk about processes of *reflective inquiries*. They might bring to the forefront the ways in which mathematics is involved in all kinds of social affairs, and question the glorifications of mathematics. Reflective inquiries are important for bringing mathematics out of an ethical vacuum. I present examples of how students can be engaged in reflective inquiries. Such inquiries might concern the nature of mathematical truth and oppose absolutist positions. They might concern socio-political issues and reveal how mathematics can be a means for identifying cases of social injustice. Reflective inquiries never result in definite conclusions. They are of an aporetic nature.

Keywords Critique of mathematics · Reflective inquiry · Mathematical practice · Social justice · Infinity · Gini index · Economic inequality · Bias index · Structural racism · Aporism

Students can learn mathematics, but they can also learn *about* mathematics. With the latter kind of learning in mind, I talk about *reflective inquiries*.[1] Such inquiries might bring to the forefront how mathematics is involved in all forms of social affairs. They might help us to question the ideology of certainty as well as the glorification of mathematics. Reflective inquiries are important for bringing mathematics out of an ethical vacuum. Any mathematical practice can engage in reflective inquiries, and in this way oppose the banality of expertise.

In this chapter I will illustrate how reflective inquiries can unfolded in educational practices. Such inquiries can concern mathematical notions, conceptions, and proofs. They can concern all forms of mathematical performatives. They can concern how

[1] In previous publications, I have used "reflective knowing" when having such philosophic issues in mind. Kennedy (2023) introduces the notion of "philosophic inquiry" with a similar interpretation.

O. Skovsmose, *Critical Philosophy of Mathematics*, Advances in Mathematics Education, https://doi.org/10.1007/978-3-031-71375-0_16

mathematics can be brought into action as a means for identifying, as well as for hiding, cases of social injustice.

Inspiration for how to invite students to engage in reflective inquiries might be found in the dialogic theory of learning mathematics that I presented in *Critical Mathematics Education* (Skovsmose, 2023, Chapter 14). This theory of learning mathematics is simultaneously a theory of learning *about* mathematics. It captures *learning of mathematics* and *learning about mathematics* as integral parts of the same process. The dialogic theory of learning mathematics is characterised by six learning interacts: *getting in contact*, *exploring*, *positioning*, *foregrounding*, *externalising*, and *doubting*. I could have talked about teaching-learning interacts, but I do not contrast learning and teaching; I see them both as features of the same process of interaction.

16.1 Even, Odd, or Both?

Deborah Ball (1993) presented an episode from a mathematics classroom, which in an exemplary way illustrates what it could mean to open up a mathematical notion to reflective inquiries. The students, all around 9 years old, were introduced to the idea of a number being even or odd. This seems to be a straightforward classification, and there does not appear to be much space for questioning it.

However, Sean, one of the students, mentioned that he found that some numbers could be both odd and even. Ball did not discard this possibility, which could have been done easily by repeating the definitions of a number being even and a number being odd. Instead, Ball asked Sean to explain what he had in mind, and in doing so Ball demonstrated what it means to respect students' thinking. Sean found that 6 was an example of a number being both odd and even, and Ball invited him to explain why. Sean's reason was that when one makes a block out of 6, then the length of one side would be 2, while the length of the other side would be 3. The length of one side would be an even number, while the length of the other side would be an odd number. The number 6 could also be blocked as a 1 times 6 rectangle, again showing that it is both even and odd. According to Sean's conception, 4 will be an even number, and 9 an odd number, while 6 would be both even and odd.

Many things could be explored with respect to Sean's proposal. Are we dealing with a true classification of the natural numbers? What about the number 12, for instance?

The process that Ball described was rich in possibilities for exploration. *Exploring* is a learning interact that includes activities such as *presenting*, *listening*, *advocating*, and *questioning*. I see *presenting* as an integral component of the whole process of learning, and Ball invited Sean to present his reasons for some numbers being both even an odd. *Listening* is an important learning interact. Students must listen to each other and to the teacher, and the teacher must listen to the students. Ball listened to Sean, and tried to grasp his perspectives. Mathematical proof is an example of *advocating*, which refers to the process of arguing for a position. Ball invited Sean to advocate for the claim that he made. *Questioning* can often make up part of a teacher's control and evaluation of the students, but most often this is not genuine questioning but rather a control procedure, as the teacher asks about something that he or she knows in advance. By referring to questioning here, I have in mind a concern for acquiring some new insight and for driving an inquiry process forward. Sean's proposal was questioned, which contributed to further exploration.

In the mathematics classroom, most often mathematical concepts are presented as given and not as issues for discussion. Mathematical definitions are to be memorised. This maxim is part of classroom absolutism that is haunting mathematics education. A teacher can challenge this absolutism by showing that mathematical concepts are indefinite, and by being ready to consider possible alternatives to given definitions. Even if there appears to be no space for alternative interpretation of a concept, it is important not to rule out this possibility.

Many classic epistemologies contrast knowing and doubting, but I do not follow this. One can think of doubt in psychological terms, and a person might be in doubt about something or other. One can also consider doubt to be an epistemic notion that is a part of knowledge construction. I do not consider doubting and knowing as contradictory epistemic attitudes, but as being features of a process of obtaining further insight. It is important to be ready to doubt any kind of assumed knowledge. The complex processes of coming to know something include equally complex processes of coming to doubt.

Doubting is a crucial component of a dialogic theory of learning mathematics, and I see *doubting* as an important learning interact. The definition of even and odd numbers appears to be untouchable by any form of doubt, and normally students will experience mathematics as being immune from any form of doubt, which contributes to the formation of the ideology of certainty. By making space for Sean's suggestion, Ball showed that doubting mathematical definitions is possible. For a critical philosophy of mathematics as well as for a critical mathematics education, it is important to confront any version of the ideology of certainty, and one way of doing so is to recognise doubt as a learning interact.

16.2 Exactly 180°?

Long ago I was teaching mathematics at an evening school for young people and adults in Denmark who had left school early and now wanted to complete their basic education. One day I asked the students, working in groups, to draw a triangle, which could be of any size and shape. I asked them to measure the angles of the triangles and the results from each group were written on the blackboard.

We looked at the numbers, and I asked what they thought about them. There came several comments, and to my surprise I realised that apparently nobody knew in advance that the sum of the angles would be 180°. I did not tell them, but asked them to calculate the sum of the angles for each of the triangles. The results surprised the students, but one of the groups suggested that it might be possible to draw a triangle with a sum bigger than 180° if it was a huge triangle. We moved the tables and chairs away, and drew a triangle on the floor, but also in this case the sum was 180°.

The students experienced that they had discovered a kind of mathematical secret. What could be the reason that the angle sums were always the same? I asked the groups to look for an explanation, and I gave them a good amount of time to do so. Then presentations from each group followed, and a surprising variety of arguments were presented. The groups, however, did not feel convinced that the presented arguments were adequate. Would the result *always* be 180°? And *exactly* 180°? Finally, when asked, I presented a regular proof.

At that time, I thought of the activity as a way for students to learn mathematics. However, the activity can also be considered a reflective inquiry. The way the mathematical properties of a triangle were explored gave the students not only a mathematical insight, but also an insight about mathematics. They recognised that empirical truths might be different from mathematical truths, which emerge with a particular degree of necessity and exactness.

One can imagine continuations of the triangle project. I could bring handballs, footballs, basketballs, and some big bouncy balls to the classroom and ask each group to choose a ball and to draw triangles on the ball. The triangles could be of any shape and size. The students could calculate the sum of the angles for each of their triangles and again write the results on the blackboard. What do they think about this? Do the results seem to depend on the size of the triangle? Or on the size of the ball? Instead of balls I could have brought a couple of watermelons to the classroom: a big one and a small one. Then the students could actually cut out triangles and measure the angles while afterwards enjoying eating watermelon.

Through such explorations, the students might gain deeper insight into the nature of mathematical truths. That the sum of the angles in a triangle is 180° is a mathematical necessity, but this necessity is relative to a specific context. If we enter a different contextualisation, other mathematical necessities might appear. That students get opportunities for reflecting and wondering is a condition for reflective inquiries to unfold in a classroom context.

Getting in contact is important for establishing cooperative processes, and consequently I have also incorporated getting in contact as a learning interact in the dialogic theory of learning mathematics.[2] To establish a dialogic relationship presupposes that one is concerned about the views of the other, and getting in contact refers to emotional elements stimulating such concerns. I was not aware of this during the project, as it took place before I got engaged in research in mathematics education. Yet, I was fully aware of the importance of creating a good atmosphere in the classroom and ensuring that everybody enjoyed being together and working together. This was crucial to give our explorations an energetic drive.

The explorations include several forms of *presentations*. The groups presented the results of their calculations and arguments for the sum of the angles being 180° to

[2]Getting in contact played an important role when Alrø and I formulated the Inquiry Cooperation Model (Alrø and Skovsmose, 2004).

each other. The students *listened* to each other, they listened to me, and I listened to them. The exploration was rich in *advocating*. When the students experienced that the sum of the angles appeared always to be 180°, it became urgent to grasp why it could be so. The students' suggested answers were forms of advocating, some being rather solid while others were weak, if not fallacious. The proof that I presented in the end appeared to still be an example of advocating. The whole process of exploration was peppered with *questioning*, both from the students and from me, driving the process forward.

16.3 Is Infinity a Number?

Nadia Stoyanova Kennedy (2007) presented an episode from a classroom where the students explored the notion of infinity.[3] The exploration was initiated by a comparison of the set of natural {1, 2, 3, 4, ... } and the set of even natural numbers {2, 4, 6, 8, ...}. After much discussion, three positions among the students crystallised: (1) the set of integers has the double number of elements compared to the set to even integers; (2) the two sets have the same number of elements; and (3) a comparison of the number of elements in the two sets is impossible. Each position had strong advocates, but nobody managed to convince the others.

The situation appeared to have stagnated until one of the students raised the question: Is infinity a number? This question gave a new drive to the inquiries:

> *Chas*: I think when people say infinity that means that they were too tired making more numbers, and they were too lazy to name any more numbers, and ... and also there would have been a lot of work.
> *Voices*: How do you know?
> *Chas*: Well, nobody bothers to go to infinity.
> *Jimmy*: Infinity is not a number.
> *Chas*: Infinity is not a number, [...] it's just in there ...
> *Sally*: If numbers are going forever I guess it's a number.
> *Chas*: It's not exactly a number, that's the name of the rest of the numbers, ... (p. 306)

While Jimmy and Chas found infinity not to be a number, Sally saw it as being a number. Following Chas' idea, infinity is an indication of a process that could go on forever. It is represented by the three dots, such as when one refers to the set of natural numbers as {1, 2, 3, 4, ... }. The three dots are used when one becomes "too tired making more numbers".

> *Rush*: ... the number infinity, it's not like 1000 billion, or 1000 trillion, it's not like that and there is a whole bunch of numbers inside infinity ... (pp. 306–307)

Rush found infinity to be a number, but a particular kind of number which contains "a bunch of numbers". This idea, however, was challenged by Claire:

[3] See also Kennedy (2006, 2012).

> *Claire*: And what number comes before infinity? I don't really think that infinity is a number. Or a bunch of numbers... (p. 307)

Claire pointed out that if infinity is a number, it must have an antecedent. If infinity is located somewhere on the number line, there must be a number just before infinity. As it seems impossible to identify such an antecedent, it is consequential to state that infinity is not a number. Jimmy restated his position:

> *Jimmy*: I think infinity is not a number, we just say this way—endless.
> [...]
> *Victor*: We are making our own definitions, 'cause the dictionary is written by people and I think they may not be always right. Infinity is just numbers that are too big.
> *Jimmy*: Infinity just means expanding and expanding and expanding. It means infiniteness. (p. 307)

The students' discussion was blossoming with ideas and suggestions. Eventually, as a result of their inquiries, they reached the conclusion that the number of elements in the set of natural is, in fact, the same as the number of elements in the set of even natural numbers.

In the discussion, different kinds of knowledge, insights, and techniques were positioned next to each other. I consider *positioning* an important learning interact. In the students' observations about infinity, one finds clear examples of positioning, where the students formulated and advocated for different conceptions of the notion. Normally the mathematics classroom is not a space that invites for positioning, but rather is regulated as a narrow one-way street leading towards pre-defined mathematical ideas.

When introducing the notion of transfinite cardinal numbers, Cantor added a new dimension to the discussion of infinity. Strong and contradictory positions were assumed by Brouwer and Hilbert. While Hilbert found the notion of infinity to be essential to mathematical theory-building, Brouwer found it to be suspicious. Inspired by Brouwer, several versions of mathematical finitism were formulated, according to which infinity is not a proper mathematical entity, but rather a process where taking one more step is always possible. Mathematics includes strong positioning with respect to infinity. In the students' reflections on the question "Is infinity a number?", they came to touch upon such deep philosophic controversies.

16.4 A Fair Distribution?

It is important to address mathematics critically. It is equally important to address socio-political issues critically, and mathematics can be an important means for doing so. One can make a *critique of mathematics*, but also a *critique by means of mathematics*. Reflective inquiries may contribute to both forms of critique.

The Gini Index is a mathematical tool for expressing degrees of economic inequality. A proper definition of the index requires concepts from calculus;

however, the idea behind the index can be expressed in less formal terms. One possibility is presented in "Ten Chairs of Inequality" by Polly Kellogg (2006).[4]

In the article "Philosophical Inquiry for Critical Mathematics Education", Kennedy (2020) described a classroom activity based on Kellogg's suggestion. As a start to the activity, the students were presented with a diagram showing the distribution of wealth in the USA in 2015.[5] The diagram showed that the wealthiest 10% of the population owns approximately 77% of the total wealth in the country; the next 10% of the population owns approximately 12% of wealth, and the remaining 80% of the population owns approximately 12% of the total wealth. The students were surprised to see the massive inequality in the distribution of wealth.

After a short discussion about the figure, the students were introduced to the following illustration of the situation. Ten chairs were used to represent the total wealth in the country, and nine students and the teacher were used to represent the total population of USA. When introducing the illustration, the teacher and the nine students were sitting on ten chairs arranged in a circle. They were asked if they found this distribution of chairs and people to be fair. They did.

The students were then asked to distribute the ten chairs among the ten people in a way that reflected the distribution of wealth in the USA. Each person represented 10% of the population. The students decided that the teacher should represent the richest 10% of the population. As this group owned approximately 77% of the total wealth, the teacher was assigned eight chairs. Sam played the role of the next 10% of the population, which owned approximately 12% of the total wealth. He was assigned one chair. The remaining chair was assigned to the remaining eight persons, as 80% of the population owns approximately 12% of the wealth.

As a result of the distribution of chairs, the teacher could lie resting comfortably across eight chairs. Sam was the owner of one chair, and could sit properly on it. The remaining eight students were the owners of one chair, around which they were all standing. Kennedy (2020) mentioned that the students had expected that after the mathematical activity everything would return to the original setting: the chairs set in a circle with everyone sitting and facing each other. However, the teacher insisted that the "material context" should be maintained for the duration of the discussion that followed. This was not a comfortable situation for the chair-less students.

The question of how wealth should be distributed became urgent, both with respect to the specific situation in the classroom, and with respect to the general economic situation in the country. One question was: Should everybody have the same income? And if not, what would then be a fair distribution? Voices were raised that rich people somehow deserved their richness:

[4] See also Staples (2006). In Chapter 3 in Skovsmose (2023) I have indicated how the Gini Index is important for discussing erosions of democracy.

[5] The diagram was presented in the *Washington Post*, May 21, 2015.

> *Rod*: Rich people don't just randomly get all this money, they worked really hard for it.
> *Voices*: Not always.
> *Rod*: You look at all of these rich people—CEOs of companies and all of that... It's not easy to get huge sums of money... so why should we take away something that someone worked really, really hard for. (p. 230)

Should wealth be redistributed in order to obtain a better degree of economic equality? Should such a redistribution be forced upon rich people? Or should it rather take the form of charity?

> *Teacher*: Do I have to give you my chairs?
> *Zack*: Did you fairly earn these chairs?
> *Teacher*: I don't know.
> *Zack*: Let's just say, you did.
> *Teacher*: I worked really hard. These are my chairs. And I did it totally legally.
> *Zack*: We'd like to have some of your chairs. You should give us some, but you don't have to.
> *Teacher*: Why are you saying this?
> *Zack*: Because as far as I'm concerned, it's the right thing to do to help poor people. But if the money is fairly earned... it's not unfair to keep all the money that you earned.
> *Jill*: The person who has all those chairs fairly earned the chairs... and there is a bunch of other people who need chairs and want chairs, there is the consideration that they need to earn their own chairs, but they can't... (p. 230)

There was much to discuss. What could they consider a fair distribution of chairs? What could they consider a fair distribution of wealth? Could the "haves" be forced to give some of their fortune to the "have nots"? Should such redistribution of wealth take the form of voluntary donations? Or of obligatory tax payments? More questions were raised: Might an unfair distribution of wealth have advantages for the general economy? Might it stimulate competition? Might it also stimulate criminality?

A mathematical description is a simplification and maybe a distortion of what is claimed to be described. As with any caricature, the Gini Index highlights some particular features and ignores others in its portrayal. By means of the Gini Index, issues of economic injustice might be exposed clearly, but the index might also tend to ignore many features of economic inequality.

Hilbert and Brouwer assumed strong *epistemic positioning* with respect to the concept of infinity, and so did the students when discussing whether or not infinity is a number. *Political positioning* is a common phenomenon in public debates; it is also common in family gatherings where a variety of political ideas might be launched. Positioning could take place with respect to any kind of topic; however, not all contexts welcome political positionings, a church service and a traditional mathematics classroom being two of them. The example of "Ten Chairs of Inequality" created a rich variety of possibilities for political positioning in the classroom. For critical mathematics education, it is important to open up classroom practices to epistemic as well as to political positionings. This brings about extra resources for dialogic interaction and reflective inquiries. I find positioning to be an important learning interact.

16.5 Killing According to Colour?

In *Critical Mathematics Education* (Skovsmose, 2023), I have defined a Bias Index as being a mathematical means for investigating problems of representativity with respect to any kind of assembly.[6] In this context I use "assembly" to refer to any group of people. My aspiration was to create a mathematical tool that could be managed by secondary-school students, and which at the same time could be a means of identifying grave cases of social injustice. Projects applying the Bias Index have still not been implemented in classroom practices, so the following presentation concerns imagined educational possibilities.

Racism need not take the form of explicit racist manifestations, but can be engraved in oppressive economic structures, in obstructions in access to welfare programmes, and in restricting educational possibilities. In Brazil, one can observe a strong bias in "favour" of black people in the assembly of people put in jail, and a strong bias in favour of white people in congress.[7] By means of the Bias Index one can put numbers on such cases of structural racism.[8]

The bias in the representation in an assembly S of a group of people A compared to the representation of a group of people B, I referred to as BI_{ABS}. It is defined as:

$$BI_{\mathrm{ABS}} = \frac{A_S B}{A B_S} = \frac{\frac{A_S}{B_S}}{\frac{A}{B}}$$

We assume $A \neq 0$, $B \neq 0$, and $B_{\mathrm{S}} \neq 0$. $\mathrm{A_S}$ refers to the group of people from A that belongs to the assembly S, while A_S (put in italics) refers to the number of people in the subset $\mathrm{A_S}$. In a similar way, $\mathrm{B_S}$ refers to the group of people from B that belongs to the assembly S, while B_S refers to the number of people in the subset $\mathrm{B_S}$.

Students can be invited to explore, by means of the Bias Index, a range of phenomena with respect to representativity. In 2017 the number of people killed by the Brazilian police was 5179; in 2018 it was 6175, and in 2019 it was 6375.[9]

[6] For addressing problems of representativity, see Barros (2021).

[7] In Brazil it is common to operate with the following ethnic groups of people: *pardos, brancos, pretos, indigenas, amarelos* (coloured, white, black, indigenous, and yellow (Asian)). For official purposes people can chose which group they feel they belong to. In 2022 the distribution was the following: *pardos* (45.3%); *brancos* (43.5%), *pretos* (10.2%) *indiginas* (0.6%), and *amarelos* (0.4%). When I talk about black people in the context of Brazil, I refer to *pardos* and *pretos*. Source: Censo 2022: pela primeira vez, desde 1991, a maior parte da população do Brasil se declara parda (ibge.gov.br)

[8] For a discussion of structural racism, see Almeida (2019) and Carrijo (2023). See also Sect. 1.2 for a clarification of my use of the notion of racism, and Sect. 15.1 for comments about structural racism. In the following definition of the Bias Index, I draw directly on formulations in Chapter 2 of Skovsmose (2023), where I also presented the mathematical reasoning that led to the formulation of the index. For further clarification of the Bias Index, see the Appendix: Bias Index.

[9] Source: https://apidspace.universilab.com.br/server/api/core/bitstreams/c23c5f89-964b-4d9b-b2d1-1328976249ab/content (p. 87).

These numbers are worrying. The killings could be caused by direct confrontations between police forces and criminal organisations. Such confrontations also cause the killing of innocent people that happen to be in the vicinity of the confrontation. Stray bullets kill many. There are cases where a police officer feels threated by a person that he or she thinks looks suspicious, and in alleged self-defence shoots towards the person. There are other cases, reported on in the media, which appear to be revenge actions by police forces, maybe because a police officer has been killed in a confrontation with gangsters.

In 2017, the distribution of people killed, one way or another, by the police was 75.4% black people (*pretos* or *pardos*), 24.4% white, and 0.2% others. That year the Brazilian population was composed of 55.8% black people, 43.1% white and 1.1% others. The degree of bias between the number of black and white people in the subdivision K, killed by the police, $\mathrm{BI}_{\mathrm{BWS}}$, can be calculated as:

$$BI_{\mathrm{BWK}} = \frac{B_K W}{BW_K} = \frac{75.4 \times 43.1}{55.8 \times 24.4} = 2.39$$

This means that the proportionality B_K/W_K (between the number of black and white people killed by the police) is 2.39 times bigger than the proportionality B/W (between the number of black and white people in the population).

The degree of bias in "favour" of black people is remarkable and is in need of further explanation. One possible explanation is that the violent criminal organisations are often located in favelas, which to a large extent are populated by black people. Here, the criminal organisations recruit people for their armed operations. Another line of explanation points to the possible racist biases within the police force in its manner of acting in threatening situations. Whatever line of explanation might be most appropriate, we are dealing with a case of structural racism.[10]

In Brazil, a body camera was introduced for the police officers to wear on their uniforms while on duty. The camera could not be turned off; it would always capture what was taking place, as seen from the perspective of the police officer. The police force is composed of the Military Police and the Civil Police, and from 2020 to 2022, the Military Police in the São Paulo State gradually introduced the use of the body camera. In Fig. 16.1, one can see the number of people killed by the Military Police from 2017 to 2023. Almost all the police operations that result in killings are carried out by the Military Police.

The number of killings dropped remarkably when the use of the camera was broadly implemented. In 2023 the use of the camera was reduced, as the State budget was reduced and could no longer cover investment in new cameras.

We can calculate the Bias Index with respect to black and white people killed by the Military Police in the São Paulo State. In making these calculations, I have used

[10] For collecting the data and completing the calculations with reference to the operations of the Military Police in the São Paulo State, I have had particular support from Denival Biotto, Fabio Alves Menescucci, and Luana P. F. Oliveira. Difficulties in accessing data might be due to obstructions by the very administrative and political system of the state (see Bueno et al., 2021).

Year	Blacks (Pretos and pardos)	White	Not identified[a]	Total
2019	474	286	99	859
2020	397	237	71	705
2021	272	113	95	480
2022	154	88	72	314
2023	220	120	38	378

[a]Here "not identified" also includes people belonging to ethnicities other black or white

Fig. 16.1 Number of people killed by the Military Police in the São Paulo State. (Source: Ouvidora da Polícia do Estado de São Paulo (2022) and https://www.ssp.sp.gov.br/ouvidoria/assets/relatorios/arquivos2/Relat%C3%B3rio_Ouvidoria_2023.pdf. The "Total" includes the number of people killed by the Military Police *in service* as well as *out of service*. The number of people killed by the Military Police *out of service* for the years from 2017 to 2022 in the São Paulo. State were 235, 179, 129, 121, 120, and 126 (Fórum Basileiro de Segurança Pública, 2023). Remarkably high numbers. I have not been able to locate the number from 2018 or for any of the previous years. Nor have I been able to find information about how killings out of service are distributed between black and white people)

the numbers for total population of the São Paulo State for 2022 being: 41.0% black (*pretos* or *pardos*), 57.8% white, and 1.2% others.[11] With these numbers the Bias Index can be calculated using the formula:

$$BI_{\mathrm{BWK}} = \frac{B_k W}{B W_k}$$

Here B_k refers to the percentage of black people killed by the Military Police, W to the percentage of white people in the São Paulo State, B to the percentage of black people in the São Paulo State, and W_k to the percentage of white people killed by the Military Police. The results are shown in Fig. 16.2.

As mentioned, in 2017, the Bias Index with respect to black and white people killed by the police in the whole of Brazil was 2.39. The black-white bias in the São Paulo State appears to be of the same level. A considerable uncertainty, however, is that the ethnicity of a relatively high number of people killed was not registered. One can consider if the Bias Index indicates some changes in the black-white bias depending on the police using cameras. It could well be that more data material, better statistical methods, as well as intensive qualitative investigations are necessary.

Instead of focusing on the groups of black and white people, one can look at the groups of young people (being defined as people from 15 to 19 years old) and adults

[11] Source: Panorama do Censo 2022 (ibge.gov.br)

Year	Bias Index
2019	2.34
2020	2.36
2021	3.39
2022	2.47
2023	2.58

Fig. 16.2 The Bias Index from 2019 to 2023 with respect to black and white people killed by the Military Police in the São Paulo State

Year	Children	Young	Adults
2017	6	172	315
2018	3	146	356
2019	2	102	435
2020	1	95	372
2021	2	42	248
2022	1	34	171
2023		39	226

Fig. 16.3 The number of children, young people, and adults killed by the Military Police in the São Paulo State from 2017 to 2022. (Source: Fórum Basileiro de Segurança Públic (2023) and https://www.ssp.sp.gov.br/ouvidoria/assets/relatorios/arquivos2/Relat%C3%B3rio_Ouvidoria_2023.pdf (p. 117). I have not been able to identify the number of children killed by the police in 2023. The number of young people killed by the police in 2023 is based on some tentative approximations)

(being defined as people 20 years old or more). The number of people killed by the Military Police is shown in Fig. 16.3.

Figure 16.3 shows a considerable drop in the numbers when the camera was introduced. This applies to the group of young people as well as to the group of adults. However, we can also consider the development of bias between these two groups. For calculating the Bias Index BI_{YAK}, where Y refers to the group of young people, A to the group of adults, and K to the group killed by the Military Police, I have used the numbers from Fig. 16.4 and the information that in 2022, 6.4% of the population in the São Paulo State is between 15 and 19 years old, and 75.6% are 20 years old or more (18.0% are children).[12] Naturally these numbers change from year to year, but as the changes are small, I have not taken such changes into consideration. With these numbers, the Bias Index can be calculated using the formula:

[12] Source: Panorama do Censo 2022 (ibge.gov.br)

Year	BI_{YAK}
2017	6.45
2018	4.84
2019	2.77
2020	3.02
2021	2.00
2022	2.35
2023	2.04

Fig. 16.4 The Bias Index from 2017 to 2023 with respect to young people and adults killed by the Military Police in the São Paulo State

$$BI_{\mathrm{YAK}} = \frac{Y_k A}{Y A_k}$$

Here Y_k refers to the percentage of young people killed by the Military Police, A to the percentage of adults in the São Paulo State, Y to the percentage of young people in the São Paulo State, and A_k to the percentage of adults killed by the Military Police. The result is shown in Fig. 16.4.

For the years 2017 and 2018, the BI_{YAK} is much higher than the following years, showing that relatively more young people than adults were killed by the Military Police. For the year 2019, before the camera was introduced, the index got smaller, but the bias was still significant. For the years 2020, 2021, 2022, and 2023 the index was around 2.00. With reference to such observations, one can definitely raise the question: Does the Military Police, when not observed, kill relatively more young people than when observed?

The Bias Index is expressed in numbers, but this does not guarantee any objectivity, nor any neutrality. One can just consider the ethnic classification of the Brazilian population: these numbers do not simply reflect objective reality, but might rather tend to hide the subjective elements associated to the classification of people. One also needs to consider the subjectivity associated with the classification of people killed, and what might be the reason for the relatively high number of people "not identified".

Working with the Bias Index, students are provided with opportunities for recognising that numbers do not safeguard any statement to be beyond doubt. Calculations by means of the Bias Index do not *justify* a specific interpretation of what is being observed, but the index might help to *identify* issues that call for interpretations and further investigations. The use of the Bias Index calls for a critique of socio-political issues, as well as for a critique of mathematics itself.

It has often been emphasised that it is important that students internalise what they have learnt. Newly acquired knowledge should become integrated in order to add to funds of already acquired knowledge. I find it to be equally important that knowledge is externalised, and consequently I see *externalising* as an important learning interact. By externalising, I refer to activities that bring the acquired knowledge beyond the horizon of the individual. Students that have addressed aspects of structural racism by means of the Bias Index will have many opportunities for externalising their insight in a broader context. This could be through presentations to the whole class, but also presentations that reach beyond the walls of the classroom.

Externalising is closely related to *positioning*. When students externalise the results of a piece of project work, these results become positioned with respect to opinions and assumed knowledge of other groups of people. Such a positioning could be both epistemic and political. It could be a positioning with respect to some already assumed knowledge which might find support or might be questioned. This could be with respect to socio-political views that might be further profiled or challenged in new ways. The notion of positioning is, however, in need of extra amplification. It not only concerns political and epistemic issues, but also personal relationships. Students can position themselves in different ways in the mathematics classroom. Some might appear confident while others might try to become invisible. This has to do with their relationship with mathematics, as has been captured by the notion of mathematical identity (see Danny Martin, 2013, 2019). Jamaal Ince (2024) has developed the notion of mathematical identify further with reference Afro-American students from New Jersey. He has revealed that racist exclusion affects black students' experiences of mathematics and their auto-positioning. They often feel that mathematics is not a topic for them; they do not feel "at home" with mathematics. Such observations bring about new dimensions to the discussion of positioning.

16.6 What Is the Risk?

In the article "Integrating Social Issues with Mathematical Modelling in Pre-Service Teacher Education", Lisa Steffensen and Georgia Kasari (2023) described a study dealing with environmental issues. Their study took place at a teacher education institution in Norway, where they focused on a course in the second year of a five-year teacher education program.

One of the activities concerned an oil spill. In 2010, a serious oil-spill disaster took place in the Gulf of Mexico at the Macondo Prospect, directed by the USA and executed by BP. It is considered one of the largest environmental catastrophes in human history. The prospective teachers were shown a map indicating the area of the gulf that was covered with oil (see Fig. 16.5). They were then presented with specific questions such as: How big is the area covered by the oil spill? How might we estimate the thickness of the oil film on the sea? It was estimated that around 7,500,000,000 litres of oil were spilled, but could this estimate be verified?

The prospective teacher came from Norway, so what could be the relevance of addressing an oil-spill problem that had taken place in the Gulf of Mexico? However, we have to remember that Norway is an oil-rich nation with many interests related to the production of oil. This also includes a high awareness of the environmental risks associated with such production. In order to grasp the profundity of such

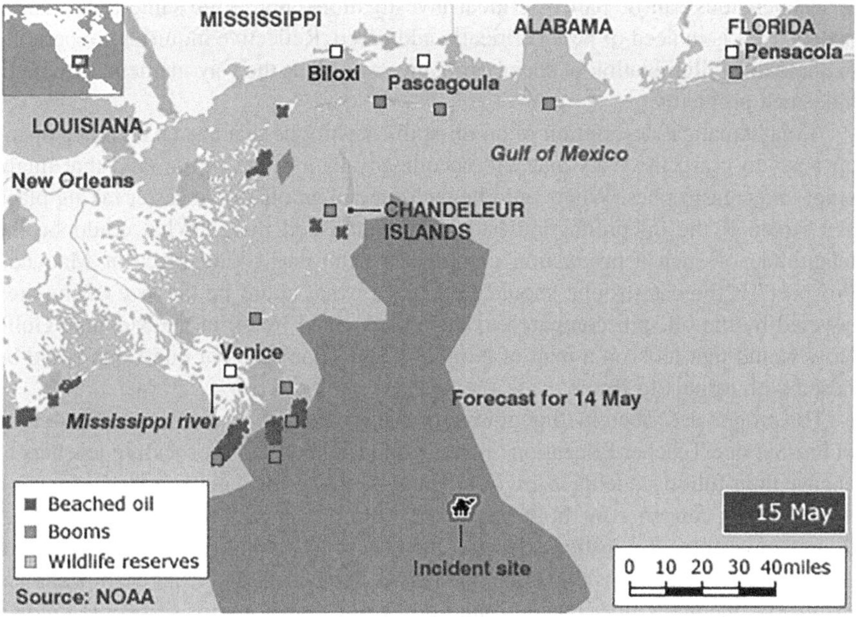

Fig. 16.5 The extension of the oil spill at the Mexican Gull in 2020. (Source: http://news.bbc.co.uk/2/hi/americas/8651333.stm)

risks, it is important to know about catastrophes that have already taken place. In Norway, strong voices have been raised against exploring any new oil fields. It has been suggested that Lofoten, a region in the northern part of Norway, should not be explored for further oil resources, nor should any of the Antarctic parts of the country. These areas are considered highly sensitive to environmental disturbances. The catastrophe that took place in the Gulf of Mexico has a direct relevance for conceptualising the risks associated with the production of oil in Norway. Brazil is also an oil-rich country, with an increasing number of off-shore oil fields, and here it becomes urgent to conceptualise the size of the risks associated with oil production. Indeed, enormous political, economic, industrial, and environmental interests are involved with respect to the outcome of any such risk estimations.

For capturing the extent of an oil-spill catastrophe, mathematical descriptions can be helpful. But mathematics is neither objective nor neutral. When using mathematics to provide a description, it is important to remember that such a description does not portray an objective piece of reality as such, but imposes an interpretation of what is being described. A mathematical description might capture something while simultaneously ignoring other things. It might grasp the extension and the dynamics of the oil spill, but overlook its impact on the life conditions of different groups of people, which was devastating for many. A mathematical description is performative. It formats the thing that is being described. It formats interpretations as well as actions.

Mathematics can be part of critical investigations, but at the same time mathematics itself is in need of being critically addressed. Reflective inquiries concern the mathematical illumination of social problems, as well as the way mathematics might hide such problems.

A mathematical description of an oil-spill catastrophe that has taken place opens up ways to grasp the risks that will accompany us into the future, and that might erupt into catastrophes. What could be the scope of an oil-spill disaster taking place at Lofoten if the oil production had been established there? What could be the magnitude of such a production compared to the one taking place at Macondo Prospect? If the catastrophe should take place, what could be the size of the area covered by the oil spill compared to the area covered by oil in the Mexican Gulf? How would that look on a map of Norway? The same kinds of questions could be raised with respect to Brazil.

The project described in "Integrating Social Issues with Mathematical Modelling in Pre-Service Teacher Education" had the aim of preparing prospective teachers to engage their future students in environmental issues. Critical mathematics education is guided by concerns for both social and environmental justice. Amongst other things, environmental justice concerns the access to natural resources that is provided for different groups of people. Resources like petrol, food, and clean water are extremely unequally distributed around the world. Some groups of people enjoy a much greater amount of natural resources than other groups. The risks associated with the extraction of natural resources are likewise distributed with a brutal degree of inequality. The deep inequalities with respect to access to natural resources and the expositions to risks are characteristics of the globalised world today, and these inequalities becomes extended into our "shared" future". For critical mathematics education, it is important to address such issues.

As a learning interact, *foregrounding* may refer to students' recognition of their own prospects in life, maybe in terms of possibilities for further education or future job opportunities. Foregrounding might draw students' attention to possibilities of which they were not previously aware. It might as well imply that students recognise obstructions in the way of reaching out to new possibilities. Foregrounding is not an isolated individual act, but takes place in dialogue with others: students, friends, families, or teachers.

Foregrounding can also concern a whole group of students. It might concern the future living conditions for today's generation of children. It might concern whether the unequal distribution of natural resources and risks is prolonged into the future, and how environmental injustice today might grow into even greater injustice to be experienced in the future. Such observations are part of foregrounding as a collective process. I see foregrounding as being an important learning interact, bringing new dynamics to reflective inquiries.

16.7 Aporism, a Human Condition?

Plato's dialogues are normally divided into three groups: the early, the middle, and the late. The early dialogues are also called aporetic, as they do not result in any clear conclusions, but end up with uncertainties and questions. Often, at the beginning of the dialogue, one of the participants makes a suggestion for how to interpret the notion in question. Although this suggestion might appear straightforward, Socrates starts questioning it. At first, the questions Socrates poses seem simple enough, but trying to answer them turns out to be difficult. The dialogues move on, and more difficulties emerge, and when the dialogue ends, nothing seems settled. The dialogues end in an aporia. The Greek word ἄπορῐ́ᾱ can be interpreted as "lack of a route forward", "deadlock", or "perplexity".

By *aporism* I refer to the conviction that all attempts to solve deeper philosophic problems, existential life issues, and basic ethical uncertainties will end up in an aporia.[13] However, I do not see this as a problem that needs to be overcome. Aporism acknowledges that doubt, as well as knowledge, is an epistemological category, and also acknowledges that doubt and knowledge are inseparable like warm and cold water when first being mixed in a bath. Aporism refers to the conviction that aporetic situations are part of any critical endeavour.

The dialogic theory of learning mathematics embraces aporism. The learning interacts are not bringing the student into realm of certainties; such acts bring about new insight, but also new questions. Advocating does not mean insisting on a certain position, but trying out possible lines of argumentation for a positioning. A positioning is seen as a possibility among various alternatives and not as a dogmatic platform. Externalising is a way of bringing the results of an exploration into the public agenda and to expose it to further questioning. Doubt is a learning interact which is an explicit expression of aporism.

Critical mathematics education encompasses a dialogical theory of learning mathematics and an aporism. The aspiration of critical mathematics education is not to guide students into definite convictions with respect to socio-political issues, but to turn such issues into part of the educational practice. The aim is to engage students in identifying and addressing cases of social and environmental injustice, and to involve them in explorations of what justice could mean. Critical mathematics education tries to leave behind any leftovers of the ideology of certainty, whether we are dealing with mathematics content matter issues or with socio-political convictions.

A critical philosophy of mathematics opposes any form of absolutism, which has tried to elevate mathematics onto an imposing pedestal from where it can shine, properly polished and out of reach for any mucky fingerprints of doubt. The aim of showing that mathematics is indefinite with respect to concepts, proofs, topics, applications, power, and culture is to open the conception of mathematics for uncertainties and doubt. The ambition of discussing the mathematical performatives

[13] I have previously addressed the notion of aporism (Skovsmose, 1998, 2000).

of shaping, fabricating, sanctifying, splitting, forming, disguising, and articulating, is to acknowledge that mathematics bring about wonders as well as horrors. A critical philosophy of mathematics sees mathematics as being constructed in a zone of contradictions and conflicts, and this brings about an ethical challenge. Therefore, ethics becomes a dimension of a critical philosophy of mathematics.

I find that aporism permeates critical mathematics education and a critical philosophy of mathematics. It permeates any critical endeavour. I see critique as a constructive expression of concerns and hopes, as well as of doubts and uncertainties. I find that aporism captures the human condition, showing it to be similar to trying to build a boat named "Knowledge" while swimming around in a turbulent ocean of uncertainties and doubt.

For a critical philosophy of mathematics, it is important to recognise that mathematics shapes our worldviews as well as our life-worlds. Mathematics makes up a key part of the formation of the human condition. This constitutes an ethical challenge for any mathematical practice, whether in research, application, or education. A critical philosophy of mathematics faces this challenge and recognises that aporetic situations are part of human life.

References

Almeida, S. (2019). *Racismo estrutural.* Editora Jandaíra.

Alrø, H., & Skovsmose, O. (2004). *Dialogue and learning in mathematics education: Intention, reflection, critique.* Kluwer Academic Publishers.

Ball, D. L. (1993). With an eye on the mathematical horizon: Dilemmas of teaching elementary school mathematics. *Elementary School Journal, 93*(4), 373–397.

Barros, D. D. (2021). *Leitura e escrita de mundo com a matemática e a communidade LGBT+: As lutas e a representatividade de um movimento social.* Doctoral thesis, Instituto de Geociências e Ciências Exatas, Universidade Estadual Paulista.

Bueno, S., Lima, R. S., & Costa, A. T. M. (2021). Quando o estado mata: Desafios para medir os crimes contra a vide de autoria de policiais. *Sociologia, Porto Alegre, 56*, 154–183.

Carrijo, M. (2023). *"Get out of my country!": Confronting racism and xenophobia through inclusive mathematics education.* Doctoral dissertation, Universidade Estadual Paulista (Unesp).

Fórum Basiliero de Segurança Públic. (2023). *As câmeras corporais na polícia militar do Estado de São Paulo: processo de implementação e impacto nas mortes de adolescentes.* Fórum Brasileiro de Segurança Pública. https://forumseguranca.org.br/publicacoes_posts/as-cameras-corporais-na-policia-militar-do-e stado-de-sao-paulo/

Ince, J. (2024). *Belief in African Americans as learners: Connecting mathematical identities and foregrounds.* Doctoral dissertation, Universidade Estadual Paulista (Unesp).

Kellogg, P. (2006). Ten chairs of inequality. In E. Gutstein & B. Peterson (Eds.), *Reading and writing the worlds with mathematics: Towards a pedagogy for social justice* (pp. 135–139). Routledge.

Kennedy, N. S. (2006). Conceptual change and dialectical transformation. In *Proceedings of the 30th conference of the International Group of the Psychology of mathematics education* (pp. 193–200).

Kennedy, N. S. (2007). From philosophical to mathematical inquiry in the classroom. *Childhood and Philosophy, 6*(6), 289–311.

Kennedy, N. S. (2012). Lipman, Dewey, and philosophical inquiry in the mathematics classroom. *Education and Culture, 28*(2), 81–94.

Kennedy, N. S. (2020). Philosophical inquiry for critical mathematics education. *Mathematics Teaching Research Journal, 12*(2), 225–234.

Kennedy, N. S. (2023). Reflective knowing in the mathematics classroom: The potential of philosophical inquiry for critical mathematics education. In M. A. Bicudo, C. Bronislaw, M. Rosa, & M. Marciniak (Eds.), *Ongoing advancements in philosophy of mathematics education* (pp. 293–307). Springer.

Martin, D. B. (2013). Race, racial projects, and mathematics education. *Journal for Research in Mathematics Education, 44*(1), 316–333.

Martin, D. B. (2019). Equity, inclusion, and antiblackness in mathematics education. *Race, Ethnicity and Education, 22*(4), 459–478.

Ouvidora da Plolícia do Estadode São Paulo. (2022). *Relatorio Anual 2022*. Ouvidora da Plolícia do Estadode São Paulo. Relatório_Ouvidoria_2022.pdf (ssp.sp.gov.br)

Skovsmose, O. (1998). Aporism: Uncertainty about mathematics. *Zentralblatt für Didaktik der Mathematik, 98*(3), 88–94.

Skovsmose, O. (2000). Aporism and critical mathematics education. *For the Learning of Mathematics, 20*(1), 2–8.

Skovsmose, O. (2023). *Critical mathematics education*. Springer.

Staples, M. (2006). Integrals and equity. In E. Gutstein & B. Peterson (Eds.), *Reading and writing the worlds with mathematics: Towards a pedagogy for social justice* (pp. 103–110). Routledge.

Steffensen, L., & Kasari, G. (2023). Integrating societal issues with mathematical modelling in pre-service teacher education. *Education Sciences, 13*, 721. https://doi.org/10.3390/educsci1307072

Afterword

When writing this book, I have been listening to music, mostly classic jazz and bossa nova. In the Introduction I mentioned that the music needed to be instrumental: guitar, piano, sax. I could not listen to songs, to words. This would disturb my mind, filled with a quite different kind of words. However, now that the manuscript is completed, I can listen to the singers.

In 1939 Billie Holiday sang *Strange Fruit* for the first time. The text was written two years previously by Abel Meeropol as a protest against the lynching of African Americans. Such lynching was still scarily common in the South of the USA. So was the circulation of lynching postcards, showing killed black people hanging from trees. *Strange Fruit* has been seen as the initiation of the civil right movement:

> Southern trees bear a strange fruit
> Blood on the leaves and blood at the root
> Black body swinging in the Southern breeze
> Strange fruit hanging from the poplar trees
>
> Pastoral scene of the gallant South
> The bulging eyes and the twisted mouth
> Scent of magnolia, sweet and fresh
> And the sudden smell of burning flesh!
>
> Here is a fruit for the crows to pluck
> For the rain to gather, for the wind to suck
> For the sun to rot, for a tree to drop
> Here is a strange and bitter crop.

The song *O Morro Não Tem Vez* (*The Slums Aren't Given a Chance*) was released in 1962. The music was composed by Antônio Carlos Jobin, at times referred to as the father of Bossa Nova; the text was written by Vinicius de Moraes. An English translation was written by Ray Gilbert. The song expresses deep concerns, but also hope.

O. Skovsmose, *Critical Philosophy of Mathematics*, Advances in Mathematics Education, https://doi.org/10.1007/978-3-031-71375-0

The slum has no say
And what it did was already too much
But pay attention you all
When they give voice to the slum
The whole city will sing

The slum wants to make its way
The slum wants to show itself
Make way for the slums
Tamborim will speak

And one, and two and three
And a hundred, a thousand beating their drums
The slum has no say
But if they give voice to the slum
The whole city will sing

From 1964 to 1985, Brazil suffered a military dictatorship. During that period, protests were expressed through songs, music, and art in general. In 1969 Chico Buarque, composed the song *Apesar de Você* (*In Spite of You*), where the *Você* (*You*) refers to the Brazilian dictatorship. At first the dictatorial censorship did not pay attention to the song, but it became extremely popular, and in 1974 the censors banned this and any song written by Buarque. Many times I have listened to the music of *Apesar de Você* played by the trio "3 Na Bossa". Here is an English translation of parts of the text.

Today you are in charge
If you said, we agree
There's no discussion
My people today
Talk about you behind your back
And looking to the floor, you see?

[...]

In spite of you
Tomorrow will be
Another day
You'll have to endure
When the morning is born
With its poems
How will you explain yourself
When the sky gets clear
Out of the blue, with impunity
My people are gonna sing
Next to you

In spite of you
Tomorrow will be
Another day
You're actually going down
And stuff

Appendix: Bias Index

The Brazilian congress has a vast majority of white people compared to black people, even though the majority of the Brazilian population is black. This is a flagrant problem of representativity, which can be explored further in terms of numbers. My aim of formulating the Bias Index was to address problems of representativity by means of elementary mathematics that could be handled by secondary-school students.

In defining the Bias Index, one considers three subsets A, B, and S of a universal set P. The P could be the population in Brazil, A and B could be two subgroups of this population, and S could be the assembly of people in congress. The number of people in P, I refer to as P, where P is put in italics. In a similar way, I refer to the numbers of elements in A, B, and S as A, B and S. $\mathrm{A_S}$ refers to the group of people from A that belongs to the assembly S, while A_S refers to the number of people in the subset $\mathrm{A_S}$. In a similar way, $\mathrm{B_S}$ refers to the group of people from B that belongs to the assembly S, while B_S refers to the number of people in the subset $\mathrm{B_S}$.

The bias of A related to B within the subset S, I refer to as BI_{ABS}. It is defined as:

$$BI_{\mathrm{ABS}} = \frac{A_S B}{A B_S}$$

We assume $A \neq 0$ and $B_S \neq 0$. See Fig. 1.[1]

One can think of the Bias Index as a function of four variables, A, B, A_S, and B_S from $Z^+ \times Z^+ \times Z^+ \times Z^+$ to Q^+. The function has the following property:

[1]The mathematical reasoning that brought me to this formulation of the Bias Index is presented in Chap. 2 of Skovsmose (2023). For exploring the Bias Index further, I have got support from Denival Biotto.

O. Skovsmose, *Critical Philosophy of Mathematics*, Advances in Mathematics Education, https://doi.org/10.1007/978-3-031-71375-0

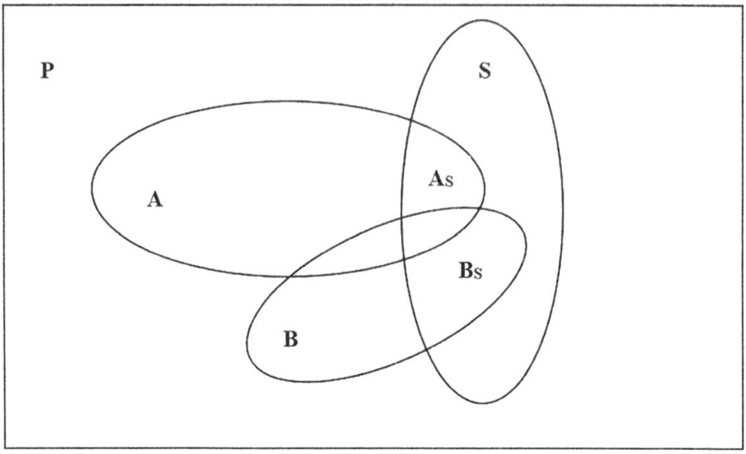

Fig. 1 The parameters of the Bias Index

$$BI_{\text{ABS}} = \frac{A_S B}{AB_S} = \frac{1}{\frac{AB_S}{A_S B_S}} = \frac{1}{BI_{\text{BAS}}}$$

meaning that the bias BI_{BAS} is the inverse of the bias BI_{ABS}.

In order to illustrate how the Bias Index is working, one can consider a simplified situation: the total population P consists of 1000 people, of which 500 are black and 500 are white. We consider an assembly S of 100 people of which *x* are black and the rest, 100—*x*, are white. We can calculate the Bias Index as:

$$BI_{\text{BWS}} = \frac{B_S W}{BW_S} = \frac{x \times 500}{500 \times (100 - x)} = \frac{x}{100 - x}$$

The graph of the function

$$BI_{\text{BWS}} = \frac{x}{100 - x}$$

is shown in Fig. 2.

The graph indicates that when x = 50, meaning that the number of black and white people in congress is the same, then the Bias Index will be $BI_{\text{BWS}} = 1.00$. Had the distribution of black and white people in the population been 750 black and 250 white and the distribution in congress been 75 black and 25 white, then the index will also be $BI_{\text{BWS}} = 1.00$.

In general, we have:

$$BI_{\text{ABS}} = \frac{A_S B}{AB_S} = \frac{\frac{A_S}{B_S}}{\frac{A}{B}}$$

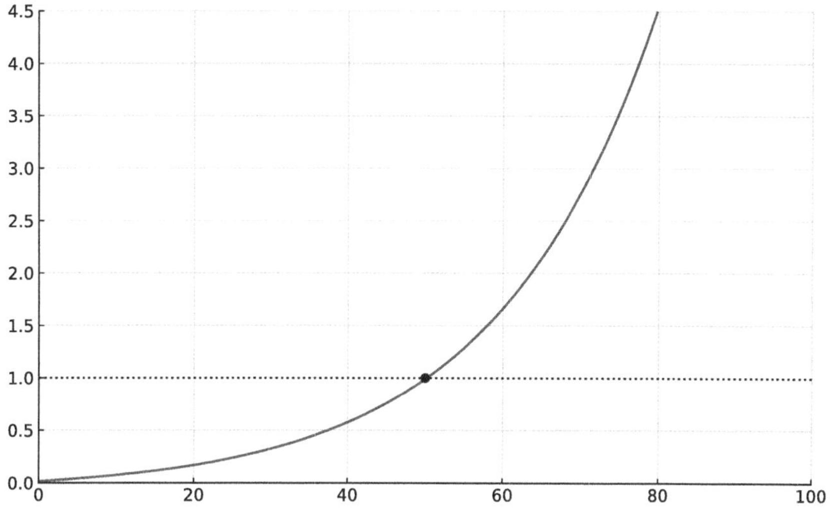

Fig. 2 The graph of the function $BI_{\mathrm{BWS}} = \frac{x}{100-x.}$

meaning that if the proportion between A_S and B_S is the same as the proportion between A and B, then $BI_{\mathrm{ABS}} = 1.00$. If the index BI_{ABS} is bigger than 1.00, then A would be overrepresented in S compared to B; and if the index BI_{ABS} is smaller than 1.00, then A would be underrepresented in S compared to B.

To argue more formally for these points, one can consider the situation where $A_S = c_1A$ where $c_1 \in]0;1]$, and $B_S = c_2B$ where $c_2 \in]0;1]$. If the proportionality between A and B within the whole population should be maintained in the congregation S, we must have $c_1 = c_2$. This means

$$BI_{\mathrm{ABS}} = \frac{A_S B}{AB_S} = \frac{Ac_1B}{Ac_2B} = \frac{c_1}{c_2} = 1$$

In other words: $BI_{\mathrm{ABS}} = 1$ if and only if $c_1 = c_{2.}$; $BI_{\mathrm{ABS}} > 1$ if and only if $c_1 > c_2$; and $BI_{\mathrm{ABS}} = 1$ if and only if $c_1 < c_2$.

Having in mind applications of the Bias Index with respect to issues of representativity of groups of people, we have considered a Bias Index to be a function of four variables from $Z^+ \times Z^+ \times Z^+ \times Z^+$ to Q^+. However, one could have many more situations in mind. One could, for instance, consider representativity of two "substances" A and B in a mixture S. The S could be a mix of yellow paint, A, and blue paint, B. In such a case it would be more adequate to consider a Bias Index to be a function of four variables from $R^+ \times R^+ \times R^+ \times R^+$ to R^+. In situations with respect to the representativity of people in an assembly, we have assumed that the ideal situation is characterised by the Bias Index being 1. However, one can imagine many situations where the ideal mix is defined in a different way, such as $c_1 = ac_2$. In this case the ideal bias would be $BI_{\text{ABS}} = a$.

A proper theory of the Bias Index could start with a definition of a function BI_{ABS} in four variables A, B, c_1, and c_2 as:

$$BI_{\text{ABS}} = \frac{Ac_1B}{Bc_2A}$$

where A $\in R^+$, B $\in R^+$, $c_1 \in]0;1]$, and $c_2 \in]0;1]$. Then one could start proving theorems like $BI_{\text{ABS}} = BI_{\text{BAS}}{}^{-1}$, and showing how properties of the function BI_{ABS} depend on relationships between the numbers c_1 and c_2. One can follow up with graphical illustrations. In this way it might be possible to engage mathematics student teachers and maybe secondary-school students in mathematical theorising, where the purpose of the theorising is possible to grasp.

Reference

Skovsmose, O. (2023). *Critical mathematics education*. Springer.

Combined Reference List

Almeida, S. (2019). *Racismo estrutural*. Editora Jandaíra.
Almeida, D. F., & Josepf, G. G. (2009). Kerala mathematics and its possible transmission to Europe. In P. Ernest, B. Greer, & B. Sriraman (Eds.), *Critical issues in mathematics education* (pp. 171–188). Information Age Publishing.
Alrø, H., & Skovsmose, O. (2004). *Dialogue and learning in mathematics education: Intention, reflection, critique*. Kluwer Academic Publishers.
Annual Register. (1800). *Annual register or a view of the history, politics, and literature for the year 1800* (Printed by William Heney, Crown Court, Aldersgate Steet).
Appel, K., & Haken, W. (1978). The four-color problem. In L. A. Steen (Ed.), *Mathematics today: Twelve informal essays* (pp. 153–180). Springer.
Arendt, H. (1977). *Eichmann in Jerusalem: A report on the banality of evil*. Penguin Books.
Arseven, A. (2015). Mathematical modelling approach in mathematics education. *Universal Journal of Educational Research, 3*(12), 973–980.
Augustine. (1961). *Confessions*. Penguin Books.
Austin, J. L. (1962). *How to do things with words*. Oxford University Press.
Avigad, J. (2008). Computers in mathematical inquiry. In P. Mancosu (Ed.), *The philosophy of mathematical practice* (pp. 302–316). Oxford University Press.
Ayer, A. J. (Ed.). (1959). *Logical positivism*. The Free Press.
Ayer, A. J. (1970). *Language, truth and logic*. Victor Gollancz.
Azzouni, J. (2005). How to nominalize formalism. *Philosophia Mathematica, 13*(2), 135–159.
Bales, K. (2012). *Disposable people: New slavery in the global economy* (3rd ed.). University of California Press.
Ball, D. L. (1993). With an eye on the mathematical horizon: Dilemmas of teaching elementary school mathematics. *Elementary School Journal, 93*(4), 373–397.
Barros, D. D. (2021). *Leitura e escrita de mundo com a matemática e a communidade LGBT+: As lutas e a representatividade de um movimento social*. Doctoral thesis, Instituto de Geociências e Ciências Exatas, Universidade Estadual Paulista.
Bauman, Z. (1989). *Modernity and the holocaust*. Polity Press and Blackwell Publishing.
Bauman, Z. (1998). *Globalization: The human consequences*. Colombia University Press.
Beck, U. (1992). *Risk society: Towards a new modernity*. Sage Publications.
Beck, U. (1999). *World risk society*. Polity Press.
Benacerraf, P., & Putnam, H. (Eds.). (1983). *Philosophy of mathematics: Selected readings* (2nd ed.). Cambridge University Press.

O. Skovsmose, *Critical Philosophy of Mathematics*, Advances in Mathematics Education, https://doi.org/10.1007/978-3-031-71375-0

Bendegem, J. P. (2020). Experiments in mathematics: Fact, fiction, or the future? In B. Sriraman (Ed.), *Handbook of the history and philosophy of mathematical practice*. Springer.

Bentham, J. (2011). *The panopticon writings*. Edited by M. Božovič. Verso Books.

Benveniste, E. (1971). *Problems in general linguistics*. University of Miami Press.

Bernal, M. (1987). *Black Athena: Afroasiatic roots of classical civilization. Volume 1: The fabrication of ancient Greece 1785–1985*. Rutgers University Press.

Bernays, P. (1983). On Platonism in mathematics. In J. van Heijenoort (Ed.), *From Frege to Gödel: A source book in mathematical logic, 1879–1931* (2nd ed., pp. 258–271). Cambridge University Press.

Betts, J. (2023). *John Harrison and the quest for longitude*. National Marine Museum.

Bishop, E. (1967). *Foundations of constructive analysis*. McGraw-Hill.

Blum, W. (1993). Mathematical modelling in mathematics education and instruction. In Breiteig (Ed.), *Teaching and learning mathematics in context* (pp. 3–14). Ellis Horwood.

Blum, W., & Niss, M. (1991). Applied mathematical problem solving, modelling applications, and link to other subjects: State, trends and issues in mathematics education. *Educational Studies in Mathematics, 22*(1), 37–68.

Blum, W., Galbraith, P. L., Henn, H.-W., & Niss, M. (2007). *Modelling in applications in mathematics education*. Springer.

Blumberg, A., & Feigl, H. (1931). Logical positivism. *The Journal of Philosophy, 18*(11), 281–296.

Boden, M. (1987). *Artificial intelligence and the natural man* (2nd ed.). MIT Press and Basic Books.

Boden, M. A. (Ed.). (2005). *The philosophy of artificial intelligence*. Oxford University Press.

Boolos, G. (1998). *Logic, logic, and logic*. Harvard University Press.

Booss-Bavnbek, B., & Høyrup, J. (Eds.). (2003). *Mathematics and war*. Birkhäuser.

Borba, M. C., & Orey, D. C. (Eds.). (2023). *Ubiratan D'Ambrosio and mathematics education: Trajectory, legacy and future*. Springer.

Borba, M., & Skovsmose, O. (1997). The ideology of certainty. *For the Learning of Mathematics, 17*(3), 17–23.

Bourbaki, N. (2004a). *Elements of mathematics: Theory of sets*. Springer.

Bourbaki, N. (2004b). *Elements of mathematics: Elementary theory*. Springer.

Bourbaki, N. (2004c). *Elements of mathematics: Set theory*. Springer.

Bourbaki, N. (2004d). Elements of mathematics: Functions of real variable. In *Springer. Selected readings* (2nd ed., pp. 202–206). Cambridge University Press.

Bourbaki, N. (2004e). *Elements of mathematics: Functions of real variable*. Springer.

Bradley, R. E., & Sandifer, E. C. (2009). *Cauchy's Cours d'analyse: An annotated translation*. Springer.

Brouwer, L. E. J. (1911). Über Abbildungen von Mannigfaltigkeiten. *Mathematische Annalen, 71*, 97–115.

Brouwer, L. E. J. (1981). In D. van Dalen (Ed.), *Brouwer's Cambridge lectures on intuitionism*. Cambridge University Press.

Brouwer, L. E. J. (1983). Intuitionism and formalism. In P. Benacerraf & H. Putnam (Eds.), *Philosophy of mathematics: Selected readings* (2nd ed., pp. 77–89). Cambridge University Press.

Brouwer, L. E. J. (1996). Life, art, and mysticism. *Notre Dame Journal of Formal Logic, 37*(3), 389–429.

Brown, G., & Yule, G. (1983). *Discourse analysis*. Cambridge University Press.

Bruno, G. (2014). *On the infinite, the universe and the worlds: Five cosmological dialogues* (Translated and introduced by Scott Gosnell). Huginn, Munnin and Co.

Bueno, S., Lima, R. S., & Costa, A. T. M. (2021). Quando o estado mata: Desafios para medir os crimes contra a vide de autoria de policiais. *Sociologia, Porto Alegre, 56*, 154–183.

Burgess, J. (2005). *Fixing Frege*. Princeton University Press.

Burton, L. (1995). Moving towards a feminist epistemology of mathematics. *Educational Studies in Mathematics, 28*, 275–291.

Bury, J. B. (1955). *The idea of progress: An inquiry into its origin and growth*. Dover Publications.
Bussell, B. (1951). *The autobiography of Bertrand Russell*. Little, Brown and Company.
Cantor, G. (1955). *Contributions to the founding of the theory of transfinite numbers* (Translated, and provided with an introduction and notes by Philop E. B. Jourdain). Dover Publications.
Carnap, R. (1937). *The logical syntax of language*. Routledge and Kegan Paul.
Carnap, R. (1959). The elimination of metaphysics through logical analysis of language. In A. J. Ayer (Ed.), *Logical positivism* (pp. 60–81). The Free Press.
Carnap, R. (1983). The logicist foundation of mathematics. In P. Benacerraf & H. Putnam (Eds.), *Philosophy of mathematics: Selected readings* (2nd ed., pp. 41–61). Cambridge University Press.
Carnap, R. (2003). *The logical structure of the world*. Open Court.
Carrijo, M. (2023). *"Get out of my country!": Confronting racism and xenophobia through inclusive mathematics education*. Doctoral dissertation, Universidade Estadual Paulista (Unesp).
Carter, J. (2019). Philosophy of mathematical practice: Motivations, themes and prospects. *Philosophia Mathematica, 27*(1), 1–32.
Carter, J. (2020). Experimental mathematics in mathematical practice. In B. Sriraman (Ed.), *Handbook of the history and philosophy of mathematical practice*. Springer.
Castells, M. (1998). *The information age: Economy, society and culture. Volume III, End of millennium*. Blackwell Publishers.
Cauchy, A.-L. (1821). *Cours d'Analyse*. De L'Imprimerie Royale.
Chandran, R. (2012). *Turing machine*. Remya Chandran.
Chronaki, A. (2010). Revisiting mathemacy: A process-reading of critical mathematics education. In H. Alrø, O. Ravn, & P. Valero (Eds.), *Critical mathematics education: Past, present and future* (pp. 31–49). Sense.
Coeckelbergh, M. (2020). *Introduction to philosophy of technology*. Oxford University Press.
Coeckelbergh, M. (2022a). *The political philosophy of AI*. Polity Press.
Coeckelbergh, M. (2022b). *Digital technologies, temporality, and the politics of co-existence*. Palgrave Macmillan.
Cohen, P. J. (1963). The independence of the continuum hypothesis. *Proceedings of the National Academy of Sciences of the United States of America, 50*(6), 1143–1148.
Cohen, P. J. (2008). *Set theory and the continuum hypothesis*. Dover Publications.
Comte, A. (2015). *A general view of positivism*. Routledge.
Copernicus, N. (1995). *One the revolutions of heavenly spheres* (C. G. Wallis, Trans.). Prometheus Books.
Corfield, D. (2003). *Towards a philosophy of real mathematics*. Cambridge University Press.
Curry, H. B. (1951). *Outlines of a formalist philosophy of mathematics*. North-Holland Publishing Company.
D'Ambrosio, U. (2024). Nonkilling mathematics? The ethics of mathematics in the final analysis. *Philosophy of Mathematics Education Journal*, (41).
Dalen, D. V. (2013). *L. E. J. Brouwer—Topologist, intuitionist, philosopher: How mathematics is rooted in life*. Springer.
Dante, A. (2003). *The divine comedy: The inferno, the purgatorio, the paradiso* (J. Ciardi, Trans.). New American Library.
Darwin, C. (1859). *On the origin of species*. John Murray.
Debarbieux, B., & Hirt, I. (Eds.). (2022). *The politics of mapping*. Wiley.
Dedekind, R. (1963). *Essays on the theory of numbers*. Dover Publications.
Dellot, B. (2017). The algorithmic workplace need not be a dystopia. *RSA Blog*. https://www.thersa.org/blog/2017/11/the-algorithmic-workplace-need-not-be-a-dystopia
Descartes, R. (1954). *The geometry of René Descartes*. (David Eugene Smith & Marcia L. Latham, Trans.). Dover Publications.
Descartes, R. (1993). *Meditations on first philosophy in focus*. Routledge.
Descartes, R. (1998). *The world and other writings*. Cambridge University Press.

Detlefsen, M. (2005). Formalism. In S. Shapiro (Ed.), *The Oxford handbook of philosophy of mathematics and logic* (pp. 236–317). Oxford University Press.

Detlefsen, M. (2008). Purity as an ideal of proof. In P. Mancosu (Ed.), *The philosophy of mathematical practice* (pp. 179–197). Oxford University Press.

Dieudonné, J. (1961). New thinking is school mathematics. In OEEC (Ed.), *New thinking in school mathematics* (pp. 31–49). OEEC.

Dieudonné, J. (1970). The work of Nicolas Bourbaki. *The American Mathematical Monthly, 77*, 134–145.

Diffie, W., & Hellman, M. (1976). New directions in cryptography. *IEEE Transaction on Information Theory, IT-22*(6), 641–654.

Dilke, C. W. (1869). *Greater Britain: A record of travel in English-speaking countries during 1866 and 1867*. Lippincott & and MacMillan and.

Doxiadis, A., Papadimitriou, C. K., Papadatos, A., & Donn, A. (2009). *Logicomix: An epic search for truth*. Bloomsbury.

Dreyfus, H. (1972). *What computers can't do: The limits of artificial intelligence*. MIT Press.

Dummett, M. (1977). *Elements of intuitionism*. Clarendon Press.

Dummett, M. (1981a). *Frege: Philosophy of language*. Duckworth.

Dummett, M. (1981b). *The interpretation of Frege's philosophy*. Duckworth.

Dummett, M. (1983). The philosophical basis of intuitionistic logic. In P. Benacerraf & H. Putnam (Eds.), *Philosophy of mathematics: Selected readings* (2nd ed., pp. 98–129). Cambridge University Press.

Dummett, M. (1991). *Frege: Philosophy of mathematics*. Duckworth.

Dunlap, A. J. with Andelman, B. *(1996). Mean business: How I save bad companies and make good companies great. Random House.*

Dusek, V. (*2006*). *Philosophy of technology: An introduction. Blackwell Publishing*.

Edmonds, D. (2020). *The murder of professor Schlick: The rise and fall of the Vienna circle*. Princeton University Press.

Eknoyan, G. (2008). Adolphe Quetelet (1796–1974): The average man and indices of obesity. *Nephrology Dialysis Transplantation, 23*(1), 47–51.

Ellul, J. (1964). *The technological society*. Vintage Books.

Engels, F. (1999). *The condition of the working class in England* (Edited with an introduction and notes by David Mclellan). Oxford University Press.

Ernest, P. (1998). *Social constructivism as a philosophy of mathematics*. State University of New York Press.

Ernest, P. (2016). A dialogue on the ethics of mathematics. *The Mathematical Intelligencer, 38*(3), 69–77.

Ernest, P. (2018). The ethics of mathematics: Is mathematics harmful? In P. Ernest (Ed.), *The philosophy of mathematics education today* (pp. 187–216). Springer.

Ernest, P. (2021). The ethics of mathematical practice. In B. Sriraman (Ed.), *Handbook of the history and philosophy of mathematical practice*. Springer.

Ernest, P. (2023). The ontological problems of mathematics and mathematics education. In M. A. V. Bicudo, B. Czarnocha, M. Rosa, & M. Marciniak (Eds.), *Ongoing advancements in philosophy of mathematics education* (pp. 3–41). Springer.

Ernest, P. (2024). *Ethics and mathematics education: The good, the bad and the ugly*. Springer.

Euler, L. (1748). *Introductio in analysin infinitorum*. Free Download, Borrow, and Streaming: Internet Archive.

Feernberg, A. (2010). *Between reason and experience: Essays on technology and modernity*. MIT Press.

Ferreirós, J. (2016). *Mathematical knowledge and the interplay of practices*. Princeton University Press.

Five Years IOWO. (1976). *Educational Studies in Mathematics, 7*(3).

Floyd, J. (2005). Wittgenstein on philosophy of logic and mathematics. In S. Shapiro (Ed.), *The Oxford handbook of philosophy of mathematics and logic* (pp. 75–128). Oxford University Press.

Flusser, V. (1999). *The shape of things: A philosophy of design*. Reaktion Books.

Fórum Basiliero de Segurança Públic. (2023). *As câmeras corporais na polícia militar do Estado de São Paulo: processo de implementação e impacto nas mortes de adolescentes*. Fórum Brasileiro de Segurança Pública. https://forumseguranca.org.br/publicacoes_posts/as-cameras-corporais-na-policia-militar-do-e stado-de-sao-paulo/

Foucault, M. (1979). *Discipline and punish: The birth of the prison*. Penguin Books.

Foucault, M. (1989). *The archaeology of knowledge*. Routledge.

Foucault, M. (2000). *Power*. The New Press.

Frege, G. (1893/1903). *Grundgesetze der Arithmetik I-II*. H. Poble.

Frege, G. (1969). Über Sinn und Bedeutung. In G. Frege (Ed.), *Funktion, Begriff, Bedeutung: Fünf Logische Studien* (pp. 40–65). Vandenhoech & Ruprecht.

Frege, G. (1978). *Die Grundlagen der Arithmetik/The foundations of arithmetic*. Blackwell.

Frege, G. (1983). Begriffsschrift: A formula language, modelled upon that of arithmetic, for pure thought. In J. van Heijenoort (Ed.), *From Frege to Gödel: A source book in mathematical logic, 1879–1931* (2nd ed., pp. 1–82). Cambridge University Press.

Freudenthal, H. (1931). Über die Enden topologischer Räume und Gruppen. *Mathematische Zeitschrift, 33*, 692–713.

Freudenthal, H. (1973). *Mathematics as an educational task*. Reidel.

Freudenthal, H. (1978). *Weeding and sowing*. Reidel.

Freudenthal, H. (1983). *Didactical phenomenology of mathematical structures*. Reidel.

Freudenthal, H. (1991). *Revisiting mathematics education: China lectures*. Kluwer.

Froude, J. A. (1886). *Oceana: Or, England and her colonies*. Longmans, Green and Co.

Fukutomi, S. (2003). Mathematics and war in Japan. In B. Booss-Bavnbek & J. Høyrup (Eds.), *Mathematics and war* (pp. 153–159). Birkhäuser.

Furuto, L. H. L., & Monkoski-Takamure, A. (2023). Voyaging beyond the horizon: An ethnomathematics legacy in Hawaii and the Pacific. In M. Borba & D. Orey (Eds.), *Ubiratan D'Ambrosio and mathematics education* (pp. 241–252). Springer.

Gabbay, M. (2010). A formalist philosophy of mathematics, Part I: Arithmetic. *Studia Logica, 96*, 219–238.

Galeano, E. (2021). *As veias abertas da América Latina*. Colecão L and PM Pocket.

Galilei, G. (1957). *Discoveries and opinions of Galileo* (Translated with an Introduction and Notes by Stillman Drake). Anchor Books.

Galilei, G. (2000). *Two new sciences* (S. Drake, Trans.). Wall and Emerson.

Galilei, G. (2001). *Dialogue concerning the two chief world systems*. The Modern Library.

Ghosh, A. (2016). *The great derangements: Climate change and the unthinkable*. Penguin books.

Giaquinto, M. (2008). Visualizing in mathematics. In P. Mancosu (Ed.), *The philosophy of mathematical practice* (pp. 22–42). Oxford University Press.

Giardino, V. (2017). Diagrammatic reasoning in mathematics. In L. Magnani & T. Bertolotti (Eds.), *Springer handbook on model-based science* (pp. 499–522). Springer.

Giardino, V. (2020). Diagrammatic proofs in mathematics: (Almost) 20 years of research. In B. Sriraman (Ed.), *Handbook of the history and philosophy of mathematical practice*. Springer.

Giddens, A. (1984). *The constitution of society: Outline of the theory of structuration*. University of California Press.

Gödel, K. (1933). Eine Interpretation des intuitionischen Aussagenkalküls. *Ergebnisse Math. Colloq, 4*, 39–40.

Gödel, K. (1940). *The consistency of the axiom of choice and of the generalized continuum hypothesis with the Axioms of Set Theory*. Princeton University Press.

Gödel, K. C. (1962). *On formally undecidable propositions of principia mathematica and related systems*. Basic Book.

Gödel, K. (1983). The completeness of the axioms of the functional calculus of logic. In J. van Heijenoort (Ed.), *From Frege to Gödel: A source book in mathematical logic, 1879–1931* (2nd ed., pp. 582–591). Cambridge University Press.

Grabiner, J. V. (1974). Is mathematical truth time-dependent? *The American Mathematical Monthly, 81*(4), 354–365.

Grassmann, H. (1861). *Lehrbuch der Arithmetikm für höhere Lehranstalten*. Verlag Adolph Enslin.

Gravemeijer, K. J., & Terwel, J. (2000). Hans Freudenthal: A mathematician on didactics and curriculum theory. *Journal of Curriculum Studies, 32*(6), 777–796.

Greer, B. (2021). Learning from history: Jens Høyrup on mathematics, education, and society. In D. Kollosche (Ed.), *Exploring new ways to connect: Proceedings of the eleventh international mathematics education and society conference, volume. 2* (pp. 487–496). Tredition.

Gutstein, E. (2006a). Math, maps, and misrepresentation. In E. Gutstein & B. Peterson (Eds.), *Reading and writing the worlds with mathematics: Towards a pedagogy for social justice* (pp. 111–123). Routledge.

Gutstein, E. (2006b). *Reading and writing the world with mathematics: Toward a pedagogy for social justice*. Routledge.

Hafner, J., & Mancosu, P. (2008). Beyond unification. In P. Mancosu (Ed.), *The philosophy of mathematical practice* (pp. 151–178). Oxford University Press.

Hahn, H. (1959). Logic, mathematics and knowledge of nature. In A. J. Ayer (Ed.), *Logical positivism* (pp. 147–161). The Free Press.

Hahn, H., Neurath, O., & Carnap, R. (n.d.-a, internet). *The scientific conception of the world: The Vienna Circle*. https:the-scientific-conception-of-the-world-the-vienna-circle(rreece.github.io)

Hahn, H., Neurath, O., & Carnap, R. (internet). *The scientific world view: The Vienna Circle*. https: the-scientific-conception-of-the-world-the-vienna-circle (rreece.github.io)

Hale, B., & Wright, C. (2005). Logicism in the twenty-first century. In S. Shapiro (Ed.), *The Oxford handbook of philosophy of mathematics and logic* (pp. 166–202). Oxford University Press.

Hall, J., & Barwell, R. (2021). The mathematical formatting of obesity in public health discourse. In A. Andersson & R. Barwell (Eds.), *Applying critical mathematics education* (pp. 210–228). Brill and Sense Publishers.

Hallett, M. (2008). Reflections on the purity of methods if Hilbert's *Grundlagen der Geometrie*. In P. Mancosu (Ed.), *The philosophy of mathematical practice* (pp. 198–255). Oxford University Press.

Hardy, G. H. (1967). *A mathematician's apology*. With a foreword by C. P. Snow. Cambridge University Press.

Hellmann, G. (2005). Structuralism. In S. Shapiro (Ed.), *The Oxford handbook of philosophy of mathematics and logic* (pp. 536–562). Oxford University Press.

Hersh, R. (1979). Some proposals for reviving the philosophy of mathematics. *Advances in Mathematics, 31*, 31–50.

Hersh, R. (1990). Mathematics and ethics. *Humanistic Mathematics Network Journal, 5*(9), 20–23.

Heyting, A. (1930). *Die formalen Regeln der intuitionistischen Logik*. Akademie der Wissenschaften. phys.-math, 42–56, 57–71, 158–169.

Heyting, A. (1971). *Intuitionism: An introduction*. North Holland Publishing.

Heijenoort, J. van. (Ed.). (1967). *From Frege to Gödel: A source book in mathematical logic, 1879–1931*. Harvard University Press.

Heijenoort, J. van. (Ed.). (1983). *From Frege to Gödel: A source book in mathematical logic, 1879–1931* (2nd ed.). Cambridge University Press.

Hilbert, D. (1902). Mathematical problems. *Bulletin of the American Mathematical Society, 8*, 437–479.

Hilbert, D. (1967). The foundations of mathematics. In J. van Heijenoort (Ed.), *From Frege to Gödel: A source book in mathematical logic, 1879–1931* (pp. 464–479). Harvard University Press.

Hilbert, D. (1983). On the infinity. In P. Benacerraf & H. Putnam (Eds.), *Philosophy of mathematics: Selected readings* (2nd ed., pp. 183–201). Cambridge University Press.

Hilbert, D. (2021). *Foundations of geometry* (E. J. Townsend, Trans.). Altus Classics.
Hilbert, D., & Ackerman, W. (1950). *Principles of mathematical logic*. Chelsea Publishing Company.
Hilbert, D., & Ackermann, W. (1950). *Principles of mathematical logic*. Chelsea.
Holford, D. (2020). The algorithmic workplace and its enactive effect on the future of professions. *Futures, 122*, 102609.
Höss, R. (1996). *Death dealer: The memories of the SS Kommandant at Auschwitz*. Da Capo Press.
Høyrup, J. (1996). The formation of a myth: Greek mathematics—Our mathematics. In G. Catherine, J. Gray, & J. Rilter (Eds.), *L'Europe mathématique, Mathematical Europe* (pp. 103–119). Editions de la Maison des sciences de l'homme.
Høyrup, J. (2020). *From Hesiod to Saussure, from Hippocrates to Jevons: An introduction to the history of scientific thought between Iran and the Atlantic*. Preprint. Max Planck Institute for the History of Science.
Hume, D. (1963). *Enquiries concerning the understanding and concerning the principles of morals*. Clarendon Press.
Hume, D. (2009). *A treatise of human nature*. The Floating Press.
Husserl, E. (1970). *The crisis of European sciences and transcendental phenomenology*. Translated with an introduction by David Car. Northwestern University Press.
Ihde, J. (1990). *Technology and lifeworld: From garden to earth*. Indiana University Press.
Ihde, J. (1998). *Philosophy technology: An introduction*. Paragon House.
Iliffle, B. (2017). *Priest of nature: The religious world of Isaac Newton*. Oxford University Press.
Ince, J. (2024). *Belief in African Americans as learners: Connecting mathematical identities and foregrounds*. Doctoral dissertation, Universidade Estadual Paulista (Unesp).
Jablonka, E., & Gellert, U. (2007). Mathematisation – Demathematisation. In U. Gellert & E. Jablonka (Eds.), *Mathematization and de-mathematization: Social, philosophical and educational ramifications* (pp. 1–18). Sense Publishers.
Jamieson, K. H., & Cappella, J. N. (2008). *Echo chamber*. Oxford University Press.
Jarrahi, M. H., & Sutherland, W. (2019). Algorithmic management and algorithmic competencies: Understanding and appropriating algorithms in gig work. In N. Taylor, C. Christian-Lamb, M. Martin, & B. Nardi (Eds.), *Information in contemporary society conference 2019* (Lecture Notes in Computer Science, vol 11420). Springer. https://doi.org/10.1007/978-3-030-15742-5_55
Jørgensen, J. (1942). *Indledning til logikken og metodelæren*. Munksgaard.
Joseph, G. G. (2000). *The crest of the peacock: The non-European roots of mathematics*. Princeton University Press.
Joseph, G. G. (2009). *Kerala mathematics: History and its possible transmission to Europe*. Kerala Council for Historical Research and Art and Humanities Research Board.
Joyce, D. E. (1998). *Euclid's elements*. http://aleph0.clarku.edu/~djoyce/java/elements/elements.html
Kaiser, G. (2020). Mathematical modelling and applications in education. In S. Lerman (Ed.), *Encyclopedia of mathematics education* (2nd ed., pp. 553–561). Springer.
Kaiser, G., & Sriraman, B. (2006). A global survey of international perspectives on modelling in mathematics education. *ZDM, 38*(3), 302–310.
Kant, I. (1973). *Critique of pure reason*. (Norman Kemp Smith, Trans.). The MacMillan Press.
Kapp, E. (2018). *Elements of a philosophy of technology* (Edited by Jeffrey West Kirkewood and Leif Weatherby, translated by Lauren K. Wolfe, afterword by Siegfried Zielinski). University of Minnesota Press.
Kellogg, P. (2006). Ten chairs of inequality. In E. Gutstein & B. Peterson (Eds.), *Reading and writing the worlds with mathematics: Towards a pedagogy for social justice* (pp. 135–139). Routledge.
Kellogg, R. B., Li, T. Y., & Yorke, J. A. (1976). A constructive proof of the Brouwer fixed-point theorem and computational results. *SIAM Journal on Numerical Analysis, 13*(4), 473–483.

Kellogg, K. C., Valentine, M., & Christin, A. (2020). Algorithms at work: The new contested terrain of control. *The Academy of Management Annals, 14*(1), 366–410. https://doi.org/10.5465/annals.2018.0174
Kennedy, N. S. (2006). Conceptual change and dialectical transformation. In *Proceedings of the 30th conference of the International Group of the Psychology of mathematics education* (pp. 193–200).
Kennedy, N. S. (2007). From philosophical to mathematical inquiry in the classroom. *Childhood and Philosophy, 6*(6), 289–311.
Kennedy, N. S. (2012). Lipman, Dewey, and philosophical inquiry in the mathematics classroom. *Education and Culture, 28*(2), 81–94.
Kennedy, N. S. (2020). Philosophical inquiry for critical mathematics education. *Mathematics Teaching Research Journal, 12*(2), 225–234.
Kennedy, N. S. (2023). Reflective knowing in the mathematics classroom: The potential of philosophical inquiry for critical mathematics education. In M. A. Bicudo, C. Bronislaw, M. Rosa, & M. Marciniak (Eds.), *Ongoing advancements in philosophy of mathematics education* (pp. 293–307). Springer.
Kissell, R. (2020). *Algorithmic trading methods: Applications using advanced statistics, optimization, and machine learning techniques*. Academic.
Kitcher, P. (1983). *The nature of mathematical knowledge*. Oxford University Press.
Kjeldsen, T. H. (2003). New mathematical disciplines and research in the wake of the World War II. In B. Booss-Bavnbek & J. Høyrup (Eds.), *Mathematics and war* (pp. 126–152). Birkhäuser.
Klagge, J. C. (2020). *Tractatus in context: The essential background for appreciating Wittgenstein's Tractatus Logico-Philosophicus*. Routledge.
Kleene, S. C. (1971). *Introduction to metamathematics*. Wolters-Noordhoff Publishing.
Klein, F. (2016a). *Elementary mathematics from an advanced standpoint. Volume I: Arithmetic, algebra, analysis*. Springer.
Klein, F. (2016b). *Elementary mathematics from an advanced standpoint. Volume II: Geometry*. Springer.
Klein, F. (2016c). *Elementary mathematics from an advanced standpoint. Volume III: Precision mathematics and approximation mathematics*. Springer.
Kleiner, I. (1989). Evolution of the function concept: A brief survey. *The College Mathematics Journal, 20*(4), 282–300.
Kline, M. (1953). *Mathematics in Western culture*. Oxford University Press.
Kollosche, D., Steflitsch, D., & Deweis-Weidlinger, K. M. (in progress). Gender, mathematics, and mathematics education. In B. Greer, D. Kollosche, & O. Skovsmose (Eds.), *Breaking images: Iconoclastic analyses of mathematics and its education*. Open Book Publishers.
Kolmogorov, A. N. (1967). On the principle of excluded middle. In J. van Heijenoort (Ed.), *From Frege to Gödel: A source book in mathematical logic, 1879–1931* (pp. 414–439). Harvard University Press.
Körner, S. (1968). *The philosophy of mathematics*. Hutchinson University Library.
Kreisel, G. (1983). Hilbert's programme. In P. Benacerraf & H. Putnam (Eds.), *Philosophy of mathematics: Selected readings* (pp. 207–238). Second Edition. Cambridge University Press.
Kuhn, T. (1962). *The structure of scientific revolutions*. Chicago University Press.
Lakatos, I. (1976). *Proofs and refutations*. Cambridge University Press.
Lange, J. de, Huntley, I., Keitel, C., & Niss, M. (Eds.). (1993). *Innovation in math education by modelling and applications*. Ellis Horwood.
Latour, B. (1993). *We have never been modern*. Harvard University Press.
Leibniz, G. (1703). *Explanation of binary arithmetic*. http://www.leibniz-translations.com/binary.htm.
Levi, P. (1987). If this is a man and the truce. *Abacus*.
Locke, J. (1997). *An essay concerning human understanding*. Penguin Books.
Lorenzen, P. (1971). *Differential and integral: A constructive introduction to classical analysis*. University of Texas Press.

Lützen, J. (1978). Funktionsbegrebets udvikling fra Euler til Dirichlet. *Nordisk Matematisk Tidsskrift, 25/26*(1), 5–32.
Lyotard, J.-F. (1984). *The postmodern condition: A report on knowledge*. Manchester University Press.
MacKenzie, D. (2007). *Mathematics of climate change: A new discipline for an uncertain century*. Mathematical Sciences Research Institute. http://library.msri.org/msri/MathClimate.pdf.
Maddy, P. (1997). *Naturalism in mathematics*. Oxford University Press.
Maddy, P. (2005). Three forms of naturalism. In S. Shapiro (Ed.), *The Oxford handbook of philosophy of mathematics and logic* (pp. 437–459). Oxford University Press.
Madler, B. (2016). *An American genocide: The United States and the California Indian catastrophe, 1846–1873*. Yale University Press.
Mancosu, P. (Ed.). (2008). *The philosophy of mathematical practice*. Oxford University Press.
Manders, K. (2008). Diagram-based geometric practice. In P. Mancosu (Ed.), *The philosophy of mathematical practice* (pp. 65–79). Oxford University Press.
Marcuse, H. (1964). *One-dimensional man*. Beacon Press.
Martin, D. B. (2013). Race, racial projects, and mathematics education. *Journal for Research in Mathematics Education, 44*(1), 316–333.
Martin, D. B. (2019). Equity, inclusion, and antiblackness in mathematics education. *Race, Ethnicity and Education, 22*(4), 459–478.
Martin-Löf, P. (1968). *Notes on constructive analysis*. Almquist and Wiksell.
Martin-Löf, P. (1982). Constructive mathematics and computer programming. In L. J. Cohen, J. Los, H. Pfeiffer, & K.-P. Podewski (Eds.), *Logic, methodology and philosophy of sciences VI*. North-Holland Publishing Company.
Marx, K. (1990). *Capital: A critique of political economy, Volume I* (Introduced by Ernest Mandel. Translated by Ben Fowkes). Penguin Books.
Marx, K. (1991). *Capital: A critique of political economy, Volume III* (Introduced by Ernest Mandel. Translated by David Fernbach). Penguin Books.
Marx, K. (1992). *Capital*: A *critique of political economy, Volume II* (Introduced by Ernest Mandel. Translated by David Fernbach). Penguin Books.
Marx, K., & Engels, F. (2010). *Manifesto of the Communist Party*. Marxist Internet Archive.
McCulloch, W. S., & Pitts, W. H. (1990). A logical calculus of the ideas immanent in nervous systems. In M. A. Boden (Ed.), *The philosophy of artificial intelligence* (pp. 22–39). Oxford University Press.
McEvoy, M. (2020). Revising "The New 4CT Problem". In B. Sriraman (Ed.), *Handbook of the history and philosophy of mathematical practice*. Springer.
McGuire, L. (2020). Feminist theories informing mathematical practice. In B. Sriraman (Ed.), *Handbook of the mathematics of the arts and sciences* (pp. 1–23). Springer.
McLuhan, M. (1962). *The Gutenberg galaxy: The making of typographic man*. University of Toronto Press.
McLuhan, M. (2001). *Understanding media: The extension of man*. Routledge.
Mill, J. S. (1970). *A system of logic*. Longman.
Monna, A. F. (1972). The concept of function in the 19th and 20 centuries, in particular with regard to the discussions between Baire, Borel and Lebesgue. *Archive for History of Exact Sciences, 9*, 57–84.
Mukhopadhya, S., & Greer, B. (in progress). Bringing an ethnomathematical perspective into classrooms. In B. Greer, D. Kollosche, & O. Skovsmose (Eds.), *Breaking images: Iconoclastic analyses of mathematics and its education*. Open Book Publishers.
Newton, I. (1999). *The principia: Mathematical principles of natural philosophy* (B. Cohen & A. Whitman, Trans.). University of California Press.
Nisbet, R. A. (1980). *History of the idea of progress*. Basic Books.
Niss, M. (1989). Aims and scope of applications and modelling in mathematics curricula. In W. Blum et al. (Eds.), *Applications and modelling in learning and teaching mathematics* (pp. 22–31). Ellis Horwood.

Niss, M., Blum, W., & Huntley, I. (Eds.). (1991). *Teaching of mathematical modelling and applications*. Ellis Horwood.
O'Connor, T., & Sandis, C. (2010). *A companion to the philosophy of action*. Wiley-Blackwell.
O'Neil, C. (2016). *Weapons of math destruction: How big data increase inequality and threaten democracy*. Broadway Books.
OECD. (1961). *Synopsis of modern secondary school mathematics*. OECD.
OEEC. (1961). *New thinking in school mathematics*. OEEC.
Orwell, G. (1987). *Nineteen Eighty-Four*. Penguin Books.
Ouvidora da Plolícia do Estadode São Paulo. (2022). *Relatorio Anual 2022*. Ouvidora da Plolícia do Estadode São Paulo. Relatório_Ouvidoria_2022.pdf (ssp.sp.gov.br)
Pambuccian, V. (2000). Brouwer's intuitionism. In B. Sriraman (Ed.), *Handbook of the history and philosophy of mathematical practice*. Springer.
Parra, A. I. S. (2018). *Curupira's walk: Prowling ethnomathematics theory through decoloniality*. Doctoral dissertation, Aalborg University Press.
Parra, A. (in progress). A performative and relational ethnomathematics. In B. Greer, D. Kollosche, & O. Skovsmose (Eds.), *Breaking images: Iconoclastic analyses of mathematics and its education*. Open Book Publishers.
Pasch, M. (1912). *Vorlesungen über neuere Geometrie*. B.G. Teubner.
Paul, S. (2020). *Philosophy of action: A contemporary introduction*. Routledge.
Peano, G. (1983). The principles of arithmetic, presented by a new method. In J. van Heijenoort (Ed.), *From Frege to Gödel: A source book in mathematical logic, 1879–1931* (2nd ed., pp. 83–97). Cambridge University Press.
Pelt, R. J., Ferreiro, L., & Greenbam, M. (Eds.). (2019). *Auschwitz: Not long ago*. Abbeville Press Publishers.
Plato. *The republic*. The Internet Classics Archive. Retrieved from http://classics.mit.edu/Plato/republic.7.vi.html.
Popper, K. R. (1963). *Conjectures and refutations: The growth of scientific knowledge*. Routledge and Kegan Paul.
Popper, K. R. (1965). *The logic of scientific discovery*. Harper and Row.
Powell, A. B., & Frankenstein, M. (1997). *Ethnomathematics: Challenging eurocentrism in mathematics education*. SUNY Press.
Ptolemy, C. (2011). *Geography* (E. L. Stevenson, Trans.). Dover Publications.
Ptolemy, C. (2014). *The almagest: Instruction to the mathematics of the heavens* (Selections translated by Bruce M. Perry. Edited with notes by William H. Donahue). Green Lion Press.
Rakus-Anderson, E. (2003). The brains behind the Enigma code breaking before the World War II. In B. Booss-Bavnbek & J. Høyrup (Eds.), *Mathematics and war* (pp. 83–102). Birkhäuser.
Ravn, O., & Skovsmose, O. (2019). *Connecting humans to equations: A reinterpretation of the philosophy of mathematics*. Springer.
Rawls, J. (1999). *A theory of justice*. Oxford University Press.
Rayo, A. (2005). Logicism reconsidered. In S. Shapiro (Ed.), *The Oxford handbook of philosophy of mathematics and logic* (pp. 203–235). Oxford University Press.
Reid, C. (1995). *Hilbert*. Springer.
Robinson, A. (1996). *Non-standard analysis*. (Revised ed.). Princeton University Press.
Rosa, M., D'Ambrosio, U., Orey, D. C., Shirley, L., Alangui, W. V., Palhares, P., & Gavarrete, M. E. (Eds.). (2016). *Current and future perspectives of ethnomathematics as a program*. Springer.
Rosenberg, O. (2022). A Gypsy in Auschwitz. *Monoray*.
Rumfitt, I. (1999). Frege's logicism. *Proceedings of the Aristotelian Society, 73*, 151–180.
Russell, B. (1905). On denoting. *Mind, 14*(56), 479–493.
Russell, B. (1919). *Introduction to mathematical philosophy*. George Allen and Unwin.
Russell, B. (1951). *The autobiography of Bertrand Russell, 1872–1914*. Little, Brown and Company.

Russell, B. (1983). Mathematical logic as based on the theory of types. In J. van Heijenoort (Ed.), *From Frege to Gödel: A source book in mathematical logic, 1879–1931* (2nd ed., pp. 150–182). Cambridge University Press.
Said, E. (1979). *Orientalism*. Vintage books.
Sapir, E. (1929). The status of linguistic as a science. *Language, 5*(4), 207–214.
Sartre, J.-P. (1989). *Being and nothingness*. Translated by Hazel E. Barnes; introduction by Mary Warnock. Routledge.
Sassen, S. (1991). *The global city: New York, London, Tokyo*. Princeton University Press.
Sauvy, A. (1952, August 14). Trois Mondes, Une Planète. *L'Observateur*.
Schaaf, W. L. (1930). Mathematics and world history. *The Mathematics Teacher, 23*(8), 496–503.
Schlimm, D., & Gonzáles, J. F. (2020). Mathematical experiments on paper and computer. In B. Sriraman (Ed.), *Handbook of the history and philosophy of mathematical practice*. Springer.
Schröder, M. R. (2009). *Number theory in science and communication: With applications in cryptography, physics, digital information, computing, and self-similarity* (5th ed.). Springer.
Schürmann, U. (2018). The separation of mathematics from reality in scientific and educational discourse. In P. Ernest (Ed.), *The philosophy of mathematics education today* (pp. 241–251). Springer.
Schürmann, U. (2023). Mathematical modelling: A philosophy of science perspective. In M. A. V. Bicudo, B. Czarnocha, M. Rosa, & M. Marciniak (Eds.), *Ongoing advancements in philosophy of mathematics education* (pp. 309–329). Springer.
Searle, J. (1969). *Speech acts*. Cambridge University Press.
Seeley, J. R. (1883). *The expansion of England*. MacMillan and Co..
Sermutlu, E. (2020). *Automata, formal language, and Turing machines*. Emre Sermutlu.
Shapiro, S. (2000). *Thinking about mathematics: The philosophy of mathematics*. Oxford University Press.
Shapiro, S. (Ed.). (2005). *The Oxford handbook of philosophy of mathematics and logic*. Oxford University Press.
Shiryaev, A. N. (2003). On the defence work of A. N. Kolmogorov during World War II. In B. Booss-Bavnbek & J. Høyrup (Eds.), *Mathematics and war* (pp. 103–107). Birkhäuser.
Shrestha, M. B. (2023). Philosophy, rigor, and axiomatics in mathematics imposed or intimately related? In M. A. V. Bicudo, B. Czarnocha, M. Rosa, & M. Marciniak (Eds.), *Ongoing advancements in philosophy of mathematics education* (pp. 409–428). Springer.
Sigmund, K. (2017). *Exact thinking in demented times: The Vienna Circle and the epic quest for the foundations of science*. Basic Books.
Sigmund-Schultze, R. (2003). Military work in mathematics 1914–1945: An attempt at an international perspective. In B. Booss-Bavnbek & J. Høyrup (Eds.), *Mathematics and war* (pp. 23–82). Birkhäuser.
Sinkevich, G. I. (2016). On the history of epsilontics. *Antiquitates Mathematicae, 10*, 183–204.
Skovsmose, O. (1994). *Towards a philosophy of critical mathematics education*. Kluwer Academic Publishers.
Skovsmose, O. (1998). Aporism: Uncertainty about mathematics. *Zentralblatt für Didaktik der Mathematik, 98*(3), 88–94.
Skovsmose, O. (2000). Aporism and critical mathematics education. *For the Learning of Mathematics, 20*(1), 2–8.
Skovsmose, O. (2005). *Travelling through education: Uncertainty, mathematics, responsibility*. Sense Publishers.
Skovsmose, O. (2012). Justice, foregrounds and possibilities. In T. Cotton (Ed.), *Towards an education for social justice: Ethics applied to education* (pp. 41–62). Peter Lang.
Skovsmose, O. (2014a). *Critique as uncertainty*. Information Age Publishing.
Skovsmose, O. (2014b). *Foregrounds: Opaque stories about learning*. Sense Publishers.
Skovsmose, O. (2018). Critical constructivism: Interpreting mathematics education for social justice. *For the Learning of Mathematics, 38*(1), 38–44.
Skovsmose, O. (2023a). *Critical mathematics education*. Springer.

Skovsmose, O. (2023b). A performative interpretation of mathematics. In M. A. V. Bicudo, B. Czarnocha, M. Rosa, & M. Marciniak (Eds.), *Ongoing advancements in philosophy of mathematics education* (pp. 269–292). Springer.

Skovsmose, O. (in progress-a). Formalism, structuralism, and the doctrine of neutrality. In B. Greer, D. Kollosche, & O. Skovsmose (Eds.), *Breaking images: Iconoclastic analyses of mathematics and its education*. Open Book Publishers.

Skovsmose, O. (in progress-b). Intuition revived. In B. Greer, D. Kollosche, & O. Skovsmose (Eds.), *Breaking images: Iconoclastic analyses of mathematics and its education*. Open Book Publishers.

Skovsmose, O. (in progress-c). A critical conception of mathematics. In B. Greer, D. Kollosche, & O. Skovsmose (Eds.), *Breaking images: Iconoclastic analyses of mathematics and its education*. Open Book Publishers.

Skovsmose, O., & Ravn, O. (2011). *Matematikfilosofi*. Systime.

Skovsmose, O., & Yasukawa, K. (2009). Formatting power of 'mathematics in a package': A challenge for social theorising? In P. Ernest, B. Greer, & B. Sriraman (Eds.), *Critical issues in mathematics education* (pp. 255–281). Information Age Publishing.

Spencer, H. (1864). *The principle of biology* (Vol. 1). Williams and Norgate.

Sriraman, B. (Ed.). (2021). *Handbook of the history and philosophy of mathematical practice*. Springer.

Sriraman, B., & Hersh, R. (2021). The history and philosophy of mathematical practice: From origin to natural historians/philosophers: A conversation. In B. Sriraman (Ed.), *Handbook of the history and philosophy of mathematical practice*. Springer.

Stadler, F. (2015). *The Vienna Circle: Studies in the origins, development, and influence of logical empiricism*. Springer.

Staples, M. (2006). Integrals and equity. In E. Gutstein & B. Peterson (Eds.), *Reading and writing the worlds with mathematics: Towards a pedagogy for social justice* (pp. 103–110). Routledge.

Steffensen, L., & Kasari, G. (2023). Integrating societal issues with mathematical modelling in pre-service teacher education. *Education Sciences, 13*, 721. https://doi.org/10.3390/educsci1307072

Stork, H. (1977). *Einführung in die Philosophic der Technik*. Wissenschaftliche Buchgesellschaft.

Swetz, F. J. (1987). *Capitalism and arithmetic: The new math of the 15th century*. Open Court Publishing Company.

Taylor, F. W. (2006). *The principles of scientific management*. Cosimo Classics.

Tiles, M. (2004). *The philosophy of set theory: An historical introduction to Cantor's paradise*. Dover Publications.

Torfing, J. (1999). *New theories of discourse: Laclau, Mouffe and Žižek*. Wiley-Blackwell.

Troelstra, A. S. (2011). History of constructivism in the 20th century. In J. Kennedy & R. Kossak (Eds.), *Set theory, arithmetic, and foundations of mathematics: Theorems, philosophies* (Lecture notes in logic, Vol. 36) (pp. 150–179). Cambridge University Press.

Turing, A. M. (1937). On computable numbers, with an application to the Entscheidungsproblem. *Proceedings of the London Mathematical Society, s2-42*(1), 230–265.

Tymoczko, T. (1979). The four-color problem and its philosophical significance. *The Journal of Philosophy, 76*(2), 57–83.

Tymoczko, T. (Ed.). (1986). *New directions in the philosophy of mathematics*. Birkhäuser.

Vesalius, A. (2007). *On the fabric of the human body: The organs of nutrition and generation*. Norman Publishing.

Vitruvius. (1960). *The ten books on architecture* (M. H. Morgan, Trans.). Harvard University Press.

Weir, A. (2016). Informal proof, formal proof and formalism. *The Review of Symbolic Logic, 9*, 23–43.

Wess, R., Klock, H., Siller, H.-S., & Greefrath, G. (2021). *Measuring professional competence for the teaching of mathematical modelling*. Springer.

Whitehead, A. (1907). *The axioms of descriptive geometry*. Cambridge University Press.

Whitehead, A. & Russell, B. (1910/1913a/1913b). Principia mathematica I-III. Cambridge University Press.

Whitt, L., & Clarke, A. W. (2019). *North American genocides: Indigenous nations, settler colonialism, and international law*. Cambridge University Press.

Whorf, B. L. (1956). *Language, thought and reality*. Edited by J. B. Carrol. MIT Press.

William, P., & Chrisman, L. (Eds.). (1994). *Colonial discourser and post-colonial theory. A Reader*. Columbia University Press.

Williams, E. (1944). *Capitalism and slavery*. University of North Carolina Press.

Wittgenstein, L. (1922). *Tractatus logico-philophicus* (With an introduction by Bertrand Russell). Routledge and Kegan Paul.

Wittgenstein, L. (1953). *Philosophical investigations*. Blackwell Publishers.

Wright, G. H. V. (1955). A biographical sketch. *The Philosophical Review, 64*(4), 527–545.

Wright, C. (1983). *Frege's conception of number as objects*. Aberdeen University Press.

Ye, F. (2011). *Strict finitism and the logic of mathematical applications*. Springer.

Zalta, E. N. (1999). Natural numbers as natural cardinals as abstract objects: A partial reconstruction of Frege's Grundgezetze in object theory. *Journal of Philosophical Logic, 28*(6), 616–660.

Zalta, E. N. (2000). Neologism? An ontological reduction of mathematics to metaphysics. *Erkenntnis, 53*(1–2), 19–65.

Zermelo, E. (1967). Investigations in the foundations of set theory I. In J. van Heijenoort (Ed.), *From Frege to Gödel: A source book in mathematical logic, 1879–1931* (pp. 199–215). Harvard University Press.

Zubofff, S. (2018). *Surveillance capitalism: The fight for human future at the new frontier of power*. Profile Books.

Name Index

O. Skovsmose, *Critical Philosophy of Mathematics*, Advances in Mathematics Education, https://doi.org/10.1007/978-3-031-71375-0

Subject Index

O. Skovsmose, *Critical Philosophy of Mathematics*, Advances in Mathematics Education, https://doi.org/10.1007/978-3-031-71375-0

SPRINGER NATURE

GPSR Compliance

The European Union's (EU) General Product Safety Regulation (GPSR) is a set of rules that requires consumer products to be safe and our obligations to ensure this.

If you have any concerns about our products, you can contact us on ProductSafety@springernature.com

In case Publisher is established outside the EU, the EU authorized representative is:

Springer Nature Customer Service Center GmbH
Europaplatz 3
69115 Heidelberg, Germany

The manufacturer's authorised representative in the EU is Springer Nature Customer Service Centre GmbH, Europaplatz 3, 69115 Heidelberg, Germany. If you have any concerns regarding our products, please contact ProductSafety@springernature.com

Printed and bound by CPI Group (UK) Ltd, Croydon, CR0 4YY
30/09/2026
02225570-0006